ogress in Mathematics 4

Edited by
J. Coates and
S. Helgason

W9-ADU-223

Complex Approximation

Proceedings, Quebec, Canada
July 3-8, 1978

Edited by
Bernard Aupetit

Birkhäuser
Boston, Basel, Stuttgart

Editor

Bernard Aupetit
Département de Mathématiques
Université Laval
Quebec G1K 7P4, Canada

Library of Congress Cataloging in Publication Data

Conference on Complex Approximation, Québec, Québec,
1978.
Complex approximation.

(Progress in mathematics; 4)
Bibliography: p.
Includes index.
1. Functions of complex variables—Congresses.
2. Functional analysis—Congresses. 3. Approximation
Theory—Congresses. I. Aupetit, Bernard, 1941–
II. Title. III. Series: Progress in mathematics
(Cambridge); 4.
QA331.C655 1978 515.9 80-12850
ISBN 3-7643-3004-X

CIP—Kurztitelaufnahme der Deutschen Bibliothek

Complex approximation:
proceedings, Quebec, Canada, July 3-8, 1978 / ed.
by Bernard Aupetit.—Boston, Basel, Stuttgart:
Birkhäuser, 1980.
(Progress in mathematics : 4)
ISBN 3-7643-3004-X

NE: Aupetit, Bernard [Hrsg.]

All rights reserved. No part of this publication may be reproduced, stored in a retrieval system, or transmitted, in any form or by any means, electronic, mechanical, photocopying, recording or otherwise, without prior permission of the copyright owner.

© Birkhauser Boston, 1980

ISBN 3-7643-3004-X

Printed in USA

To John Wermer

LIST OF PARTICIPANTS

H. ALEXANDER, University of Illinois at Chicago Circle, Chicago, U.S.A.

B. AUPETIT, Université Laval, Québec, Canada.

J. BECKER, Purdue University, West Lafayette, U.S.A.

E. BEDFORD, Princeton University, Princeton, U.S.A.

J. BRENNAN, University of Kentucky, Lexington, U.S.A.

J. BURBEA, University of Pittsburgh, Pittsburgh, U.S.A.

J. CHAUMAT, Université de Paris-Sud, Orsay, France.

A.-M. CHOLLET, Université de Paris-Sud, Orsay, France.

J.A. CIMA, University of North Carolina, Chapel Hill, U.S.A.

R. COUTURE, Université Laval, Québec, Canada.

D.G. DICKSON, University of Michigan, Ann Arbor, U.S.A.

J.E. FORNAESS, Princeton University, Princeton, U.S.A.

W.H. FUCHS, Cornell University, Ithaca, U.S.A.

C. GAUTHIER, Université Laval, Québec, Canada.

P. GAUTHIER, Université de Montréal, Canada.

J.J. GERVAIS, Université Laval, Québec, Canada.

R. GERVAIS, Collège Militaire Royal, Saint-Jean, Canada.

A. GIROUX, Université de Montréal, Montréal, Canada.

M. GOLDSTEIN, Arizona State University, Tempe, U.S.A.

G.A. HARRIS, Texas Tech. University, Lubbock, U.S.A.

L.I. HEDBERG, Université de Stockholm, Stockholm, Suède.

W. HENGARTNER, Université Laval, Québec, Canada.

A.M. KHAN, McMaster University, Hamilton, Canada.

D. KUMAGAI, Lehigh University, Bethlehem, U.S.A.

M. LABRECHE, Université de Montréal, Montréal, Canada.

M. LAVOIE, Université du Québec à Chicoutimi, Chicoutimi, Canada.

H.P. LEE, Brown University, Providence, U.S.A.

H. LORD, Université Laval, Québec, Canada.

V.P. MADAN, Red Deer College, Red Deer, Canada.

M. OUELLET, Université Laval, Québec, Canada.

W. H. OW, Michigan State University, East Lansing, U.S.A.

K. K. PURI, University of Maine, Orono, U.S.A.

L. A. RUBEL, University of Illinois, Urbana, U.S.A.

J. RUIZ FERNANDEZ DE PINEDO, Consejo Superior de investigaciones cientificas Madrid, Espagne.

S. SCHEINBERG, University of California, Irvine, U.S.A.

J.A. SIDDIQI, Université Laval, Québec, Canada

K. SRINIVASACHARYULU, Université de Montréal, Montréal, Canada.

E. L. STOUT, University of Washington, Seattle, U.S.A.

J. L. WANG, University of Alabama, University, U.S.A.

B. WEINSTOCK, University of North Carolina, Charlotte, U.S.A.

J. WERMER, Brown University, Providence, U.S.A.

TABLE OF CONTENTS

Introduction v

Invited Talks

- H. Alexander: On the area of the spectrum of an element of a uniform algebra. 3
- B. Aupetit: Sous-harmonicité et algèbres de fonctions. 13
- E. Bedford and J.-E. Fornaess: Approximation on pseudoconvex domains. 18
- J. Brennan: Point evaluations, approximation in the mean and analytic continuation. 32
- J. Chaumat: Quelques propriétés du prédual de H^∞. 47
- A.-M. Chollet: Ensembles pics de $A^\infty(D)$. 57
- W.H. Fuchs: On Chebychev approximation on several disjoint intervals. 67
- P. Gauthier: Uniform analytic approximation on unbounded sets (texte non parvenu).
- L.I. Hedberg: Approximation in L^p by harmonic functions. 75
- N. Kalton and L.A. Rubel: Gap interpolation theorems for entire functions. 77
- J.A. Siddiqi: L'approximation exponentielle dans $\mathbb{C}$ (texte non parvenu).
- E.L. Stout: Uniform approximation on certain unbounded sets in $\mathbb{C}^n$. 80
- B. Weinstock: Uniform approximation on smooth polynomially convex sets. 83

Supplementary Talks

- B. Aupetit: L'approximation entière sur les arcs allant à l'infini dans $\mathbb{C}^n$. 93
- G.A. Harris: An algebraic question related to the function for real submanifolds of $\mathbb{C}^n$. 103
- S. Scheinberg: Approximation and non-approximation on Riemann surfaces. 110

INTRODUCTION. This book contains the texts of the lectures given by the invited lecturers at the Conference on Complex Approximation held at Quebec on July 3-8, 1978. It contains also supplementary papers resulting from discussion which took place during this meeting.

The three main subjects were: approximation in $\mathbb{C}^n$ and function algebras, analytic and harmonic approximation in $\mathbb{C}$, exponential approximation and approximation in L^p.

We received substantial financial help from the National Research Council of Canada and the Ministry of Education of the Province of Quebec, and from Laval University.

In the name of all participants we sincerely thank these institutions for having made this successful meeting possible.

Bernard Aupetit

INVITED TALKS

ON THE AREA OF THE SPECTRUM OF AN ELEMENT OF A UNIFORM ALGEBRA

by

H. Alexander

INTRODUCTION. In classical function theory, the area of the image of a holomorphic function was usually computed *with multiplicity*. In [5], Alexander, Taylor and Ullman obtained an estimate for the area, *without multiplicity*, of the image of a function holomorphic in the unit disc. This had applications to function theory. Here we shall discuss an area theorem in the context of uniform algebras where an estimate for the planar area of the spectrum of an element of the uniform algebra will be obtained. The proof which we shall give will depend on a quantitative version of the classical Hartogs-Rosenthal theorem on rational approximation in the complex plane. Applied to certain polynomial algebras, this "area theorem" yields properties of analytic subvarieties of $\mathbb{C}^n$. Hartogs' theorem, the separate analyticity implies analyticity, and an analogous result of Nishino, that separate normality implies normality, are consequences.

To fix some notation, $C(X)$ will denote the Banach algebra of all continuous complex-valued functions on a compact Hausdorff space X, normed with the supremum norm. When X is a compact subset of $\mathbb{C}^n$, $C(X)$ has subalgebras $P(X)$ and $R(X)$ which are the closures in $C(X)$ of the polynomials (in the coordinates) and the rational functions, holomorphic on a neighborhood of X, respectively. The maximal ideal space of $P(X)$ can be identified with the polynomially convex hull $\hat{X}(\equiv\{p\in\mathbb{C}^n : |f(p)|\le|f|_X$ for every polynomial $f\})$ of X. Thus, for a polynomial f, the plane set

$f(\hat{X})$ is the spectrum of f, considered as an element of the Banach algebra $P(X)$. It is this set whose area we shall estimate. For the fundamentals on uniform algebras and polynomial convexity we refer to the books of Stout [11] ans Wermer [12].

AREA THEOREM. First recall the classical area formula. Let f be holomorphic in the open unit disc U with $f(0) = 0$ and Taylor series $f(z) = \sum_{n=1}^{\infty} a_n z^n$. Then, as a mapping from $\mathbb{R}^2$ to $\mathbb{R}^2$, the Jacobian determinant of f is $|f'|^2$ and so the planar area of the image of f, *counting multiplicity*, is given by

$$\int_U |f'|^2 \, dxdy = \pi \sum_{n=1}^{\infty} |a_n|^2 . \tag{1}$$

the integral being easily evaluated in polar coordinates. Now suppose that f has L^2 boundary values, also denoted by f; i.e., take f to be in the Hardy space H^2.. Then, for the measure $dm = \frac{1}{2\pi} d\theta$ on the unit circle T,

$$\int_T |f|^2 dm = \sum_{n=1}^{\infty} |a_n|^2 . \tag{2}$$

The obvious estimate in (1) and (2) gives

$$\text{area, with mult., of } f(U) \geq \pi \int_T |f|^2 dm. \tag{3}$$

EXAMPLE: Take $f(z) = z^5$. On the left side of (3) we have 5π, the area, *with multiplicity*, of the image of f, while the integral on the "right side of (3) equals π, which, in fact, is the area of the image of f *without* counting multiplicity.

The estimate (3) will be generalized as follows:

$$\text{area, } \textit{without} \text{ mult., of } f(U) \geq \pi \int_T |f|^2 dm \tag{4}$$

Observe that in the above example, (4) becomes an equality. The estimate (4) was obtained by Alexander, Taylor and Ullman [5]. Here we shall give a general version which is valid for uniform algebras; the following proof (see [2] and [3]) is not the original one of [5] .

Let A be a uniform algebra with maximal ideal space M, let $x \in M$, and let σ be a (positive) representing for x supported on M. (In applications, we usually take σ to live on the Shilov boundary.) Planar Lebesque measure will be denoted be λ below. We can now state the generalization to the setting of uniform algebras of (4) which itself is the special case of A the disc algebra, x the origin and σ the measure $\frac{1}{2\pi}d\theta$.

THEOREM 1 [3]. *Let $f \in A$ and $f(x) = 0$, then*

$$\lambda(f(M)) \geq \pi \int |f|^2 d\sigma \quad . \tag{5}$$

REMARK: Since this requires the functions to be continuous on the maximal ideal space, one can apply the theorem on discs of radius less than one and take a limit to get (4) form (5).

The proof will be based on the following quatitative form of the Hartogs-Rosenthal theorem.

THEOREM [2]. *Let K be a compact subset of the complex plane. Considering $\bar{z}$ as a function in C(K), one has the following estimate for the distance from $\bar{z}$ to the subset R(K) of C(K):*

$$dist\ (\bar{z}, R(K)) \leq \left(\frac{\lambda(K)}{\pi}\right)^{1/2} .$$

PROOF: Let ψ be a C^∞ function with compact support in the plane such that $\psi(z) \equiv \bar{z}$ on a neighborhood of K. By the generalized Cauchy integral formula,

$$\psi(z) = -\frac{1}{\pi}\iint \frac{\partial\psi}{\partial\bar{\zeta}} \frac{dudv}{\zeta - z}, \quad \zeta = u + iv \ ,$$

for all $z \in \mathbb{C}$. Restricting ψ to K and using $\partial\psi/\partial\bar{\zeta} \equiv 1$ on K we get

$$\bar{z} = -\frac{1}{\pi}\iint_K \frac{dudv}{\zeta - z} - \frac{1}{\pi}\iint_{\mathbb{C}\setminus K} \frac{\partial\psi}{\partial\bar{\zeta}} \frac{dudv}{\zeta - z}$$

for $z \in K$. Since the second integral represents a function in R(K), we have

$$\operatorname{dist}(\bar{z}, R(K)) \leq \sup_{z \in K} \left|\frac{1}{\pi}\iint_K \frac{dudv}{\zeta - z}\right| \tag{6}$$

By an elegant computation, Ahlfors and Beurling ([1], p. 106-107) have shown that the right side of (6) is dominated by $(\lambda(K)/\pi)^{1/2}$. □

REMARK: If $\lambda(K) = 0$, Lemma 2 says that $\bar{z} \in R(K)$. Then, by the Stone-Weierstrass theorem, R(K) = C(K). This is the Hartogs-Rosenthal theorem.

PROOF OF THEOREM 1: Let $\epsilon > 0$ and put $K = f(M)$. Use the lemma to obtain a rational function r(z) with poles off K such that

$$|\bar{z}-r(z)|_K < \left(\frac{\lambda(K) + \epsilon}{\pi}\right)^{\frac{1}{2}}.$$

Since r is is holomorphic on a neighborhood of the spectrum K of f, it follows from the Gelfand theory that $g = r \circ f \in A$ and $|\bar{f}-g|_M < [(\lambda(K)+\epsilon)/\pi]^{\frac{1}{2}}$. We have $|f|^2 = f(\bar{f}-g) + fg$. Since $g \in A$, we get $\int fg d\sigma = f(x) \times g(x) = 0$ and so $\int |f|^2 d\sigma = \int f(f-g)d\sigma$. Thus $\int |f|^2 d\sigma \leq |\bar{f}-g|_M \int |f| d\sigma \leq [(\lambda(K)+\epsilon)/\pi]^{\frac{1}{2}} \int |f| d\sigma$. Now letting $\epsilon \to 0$ yields

$$\int |f|^2 d\sigma \leq \left(\frac{\lambda(K)}{\pi}\right)^{\frac{1}{2}} \int |f| d\sigma \tag{7}$$

This is somewhat more general than (5) (and can be used to study the case of equality in (5)). An application of Hölder's inequality, $\int |f| d\sigma \leq (\int |f|^2 d\sigma)^{\frac{1}{2}}$, in (7) gives (5).

Our first application is to the area of an analytic variety in $\mathbb{C}^n$ of complex dimension one -" a Riemann surface with singularities". 1-varieties have a naturel Lebesque area when viewed as real two-dimensional surfaces of $\mathbb{R}^{2n} = \mathbb{C}^n$. In fact, if we put $\omega = i/2 \sum_{k=1}^{n} dz_k \wedge d\bar{z}_k$, then for any analytic 1-variety V in C^n, we have

$$\text{area}(V) = \int_V \omega , \tag{8}$$

where the right side indicates integration of the real differential 2-form ω over the real oriented 2-dimensional surface V. The verification [7] of (8) amounts to two observations: (a) (8) is clearly valid if V is linear space and (b) ω is invariant under translations and unitary transformations. Now writing $\int_V \omega = \sum_{k=1}^{n} \frac{i}{2} \iint_V dz_k \wedge d\bar{z}_k$ and observing that $\frac{i}{2} dz \wedge d\bar{z} = dx \wedge dy$ for $z = x + iy$, we see that the kth term in this sum is just the area of $z_k(V)$, counted with multiplicity, where $z_k(V)$ is the planar image of V under the kth coordinate function z_k. Thus, for any 1-variety V in $\mathbb{C}^n$,

$$\text{area}(V) = \sum_{k=1}^{n} [\text{area, with mult., of } z_k(V)] \tag{9}$$

Now suppose that V is a subvariety of a ball of radius R in $\mathbb{C}^n$ which passes through the center of the ball which we take to be the origin. Then it is known[10] that the area of V is bounded below and that the extremal case occurs when V is a linear space; namely, area $(V) \geq \pi R^2$. In view of (9), the following estimate, first obtained in [5], generalizes this.

$$\sum_{k=1}^{n} [\text{area, } \textit{without} \text{ mult., of } z_k(V)] \geq \pi R^2. \tag{10}$$

We shall prove a version of this in the more general setting of polynomial hulls.

THEOREM 3 [2]: *Let Y be a compact subset of the sphere, centered at the origin, of radius R in* $\mathbb{C}^n$ *and put* $X = \hat{Y}$. *Suppose that X contains the origin. Then*

$$\sum_{k=1}^{n} \lambda(z_k(X)) \geq \pi R^2 \tag{11}$$

REMARK: To obtain (10), fix $r < R$ and let Y be the intersection of V with the sphere of radius r. Then X is the intersection of V with the closed ball of radius r. Now apply (11) and let $r \uparrow R$. In the same way, by invoking the local maximum modulus principle, one can improve (11) by replacing X with $X \setminus Y$.

PROOF: In Theorem 1 take A to be P(X) (with maximal ideal space X) and x to be the origin. Let σ be a representing measure for the origin $(0 \in X)$ which has its support in Y. Then, as $z_k \in P(X)$, we have

$$\lambda(z_k(X)) \geq \pi \int |z_k|^2 d\sigma.$$

Summing over k gives

$$\sum_{k=1}^{n} \lambda(z_k(X)) \geq \pi \int \{\sum_{k=1}^{n} |z_k|^2\} d\sigma.$$

But $\sum_{1}^{n} |z_k|^2 \equiv R^2$ on Y and so (11) follows. □

For our second application of (5) we shall consider a one-dimensional subvariety V of the unit polydisc U in $\mathbb{C}^2$. Let z and w be the coordinate functions in $\mathbb{C}^2$.

THEOREM 4. *Let* $p = (a,b)$ *be a fixed point of V. Suppose that there exists a* $\delta > 0$ *such that* $|w| > \delta$ *on V. Then*

$$\lambda(z(V)) \geq \pi \frac{Log \frac{1}{|b|} (1-|a|)^2}{log (\frac{1}{\delta})} \tag{12}$$

REMARKS: While this estimate on the area of the **z**-projection of V is not usually best possible, asymptotically this is the case when $a = 0$ and $|b| \downarrow \delta$; for then both sides of (12) approach π. The "vertical" variety $z = a$ and its small pertubations, which have small projections into the z-plan, are ruled out by the condition $|w| > \delta$ on V.

Moreover, the strict positivity of δ is needed to force a lower bound on the area of z(V). Indeed, given $0 < b < 1$, one can construct a subvariety V of U^2 containing the point (0,b), on which $w \neq 0$, and such that z(V) has arbitrarily small area ϵ. To obtain V, let K be a compact subset of the unit disc U which is disjoint form the closed unit interval [0,1] and such that the area of U\K is less than ϵ. By Runge's theorem, there exists a polynomial g such that $|g-1| < 1/2$ on K and $g(0) = Log\ b$. Now put $f = \exp(g)$ and $V = \{(z,f(z)): z \in U\} \cap U^2$. Then V is a 1-subvariety of U^2, $(0,b) \in V$, $w \neq 0$ on V, and $\lambda(z(V)) < \epsilon$, because $|f(z)| > \sqrt{e}$ for $z \in k$ implies $z(V) \subseteq U \setminus K$.

PROOF: Without loss of generality we may assume that V extends to be a subvariety of a neighborhood of the closed unit polydisc; this is because we can work with small dilations of the original variety and take a limit. This means that $\bar{V} \cap \bar{U}^2$ is the maximal ideal space of the algebra $A = P(\bar{V} \cap \bar{U}^2)$ and that the Shilov boundary Σ of A is contained in ∂U^2. Let σ be a Jensen measure supported on Σ which represents the point p for the algebra A. Then, since the coordinate function w is an invertible element of A, we have

$$\log\left(\frac{1}{|b|}\right) = \int \log\left(\frac{1}{|w|}\right) d\sigma = \int_{\{z=1\}} \log\left(\frac{1}{|w|}\right) d\sigma \tag{13}$$

$$\leq \log\left(\frac{1}{\delta}\right) \cdot \sigma\{|z| = 1\}$$

On the other hand, by (5) applied for $f \equiv z-a$, we get

$$\pi\sigma\{|z|=1\}\ (1-|a|)^2 \le \pi \int_{\{|z|=1\}} |z-a|^2 d\sigma \tag{14}$$

$$\le \pi \int |z-a|^2\, d\sigma \le \lambda((z-a)(V)) = \lambda(z(V)).$$

Now (12) follows form (13) and (14). □

To demonstrate how these "area" results can be applied, we shall give a proof of the following theorem of Nishino.

THEOREM 5[9]: *Let F be a family of holomorphic functions on a domain* Ω *in* $\mathbb{C}^n$. *Suppose that F is normal in each variable separately; i.e., suppose that each of the families of functions on the appropriate plane domains obtained from elements of F by fixing n-1 of the coordinates is a normal family. Then F is a normal family.*

REMARK: In order to more clearly illustrate the idea, we shall take "normal" to mean "uniformly bounded on compact subsets" and take $\Omega \subseteq \mathbb{C}^2$; the method can easily be adapted to treat convergence to infinity and the $\mathbb{C}^n$ case.

PROOF: Since normality is a local property, it suffices to suppose that $\bar{U}^2 \subseteq \Omega$ and to show that F is locally bounded on some neighborhood of the origin. Arguing by contradiction, we suppose ortherwise; namely, we suppose that there is a sequence $\{f_k\} \subseteq F$ and $z_k (\in U^2) \to 0$, with $\| z_k \| < \frac{1}{2}$, such that $w_k \equiv f_k(z_k) \to \infty$. Let $V_k = \{z \in U^2 : f_k(z) = w_k\}$; V_k is a subvariety of U^2. Applying (10) to the intersection of V_k with the ball of radius $\frac{1}{2}$ centered at z_k we get

$$\lambda(z(V_k)) + \lambda(w(V_k)) \ge \pi(\tfrac{1}{2})^2.$$

Thus either $z(V_k)$ or $w(V_k)$ has area at least $\frac{1}{2}.\pi(\frac{1}{2})^2$. Without loss of generality (passing to a subsequence if necessary and relabeling) we may assume that

$$\lambda(z(V_k)) \ge \tfrac{1}{2}\pi(\tfrac{1}{2})^2 \tag{15}$$

for each k. But (15) implies, by elementary measure theory, that there is a point z_0(in fact, a set of points of positive measure) contained in $z(V_k)$ for infinitely many k. This means that $\{f_k(z_0,w)\}$ is not uniformly bounded for $|w|\le 1$ because $f_k(z_0,w) = w_k$ whenever $(z_0,w)\in V_k$. This contradicts our hypothesis that $\{f_k(z_0,w)\}$ is normal in w. □

Next let's see how the area estimates lead to a proof of Hartogs' theorem that a function of two complex variables which is holomorphic in each variable separately is holomorphic. We shall derive the following result of Hartogs' as an application of Theorem 4; the usual derivation depends of Hartogs' lemma on subharmonic functions (see Hörmander [8], p.21).

LEMMA 6 (Hartogs): *Let $f(z,w)$ be a complex valued function defined on a neighborhood of $\bar{U}^2$ which is (a) holomorphic for $w\in\bar{U}$ for each fixed $z\in U$ and (b) holomorphic on $\{(z,w);|z|<1$ and $|w|<\delta\}$ for some $\delta>0$. Then f is holomorphic in U^2.*

PROOF: For each fixed $z\in U$, write the Taylor series in w, $f(z,w) = \sum_{n=0}^{\infty} a_n(z)w^n$. By (b), the functions $a_n(z)$ are holomorphic in U. Put $f_N = \sum_{n=0}^{N} a_n(z)w^n$; f_N is holomorphic in U^2. It clearly suffices to show that the family $\{f_N\}$ is uniformly bounded on compact subsets of U^2. Arguing by contradiction, we suppose not; then there is a subsequence $\{N_j\}$ of the positive integers and a sequence $\{p_{N_j}\}$ in U^2 converging to some point of U^2 and such that $f_{N_j}(p_{N_j}) \equiv w_j \to\infty$. Let $V_j = \{(z,w)\in U^2\colon f_{N_j}(z,w) = w_j\}$. By (b), we may assume that $|w|>\delta$ on each V_j. Applying Theorem 4 to V_j (taking p_j to be p), we conclude that there is an $\eta>0$ such that $\lambda(z(V_j))>\eta$ for each j. As above, it follows that there is a point z_0 in $z(V_j)$ for infinitely many j and hence that $\{f_N(z_0,w)\}$ is not uniformly bounded for $|w|\le 1$. This contradicts (a) and the proof is complete. □

Now Hartogs' theorem can be easily deduced. The full details are in Bochner-Martin [6]; here is a sketch. It suffices to assume that f is defined on a neighborhood of $\bar{U}^2$ and separately holomorphic and to show that f is holomorphic on U^2. As $M(w) \equiv \sup\{|f(z,w)|:|z|\leq 1\}$ is finite for each $w \in U$, the Baire category theorem implies that M is uniformly bounded on some open subset of U. By a Möbius transformation in the w-variable, we may assume that f is bounded on the set $\Omega = \{(z,w):|z|\leq 1 \text{ and } |w|\leq\delta\}$. The boundedness implies that f is holomorphic on Ω (continuity follows from boundedness by Schwarz's lemma and analyticity follows from continuity by the multiple Cauchy integral formula). Now Lemma 7 can be applied to conclude that f is holomorphic in U^2.

Finally we shall mention, without proof, a few more results along the above lines. In Alexander-Osserman [4] area estimates are given for one-dimensional subvarieties of tube domains and for subvarieties of a ball not passing through the center; it is also shown that equality holds in (4) if and only if f is a constant multiple of an inner function. In [3], a version of Theorem 1 is given in which area is computed in the spherical metric. This has applications to the area in projective space of the image of a subvariety of the ball under the natural projection. One consequence is a "radial" version of Nishino's result (Theorem 5 above) on separate normality.

REFERENCES

[1] L. Ahlfors and A. Beurling, Conformal invariants and function-theoretic null-sets, *Acta Math.* 83 (1950), 101-129.

[2] H. Alexander, Projections of polynomial hulls, *J. Funct. Anal.* 13 (1973), 13-19.

[3] H. Alexander, Volumes of images of varieties in projective space and in Grassmannians, *Trans. Amer. Math. Soc.* 189 (1974), 237-249.

[4] H. Alexander and R. Osserman, Area bounds for various classes of surfaces, *Amer. J. of Math.* 97 (1975), 753-769.

[5] H. Alexander, B.A. Taylor and J. Ullman, Areas of projections of analytic sets, *Inventiones Math.* 16 (1972), 335-341.

[6] S. Bochner and W.T Martin, *Several Complex Variables*, Princeton U. Press, Princeton, 1948.

[7] G. de Rham, On currents in an analytic complex manifold, *Seminars in Analytic Functions*, vol. 1, Princeton, N. J., 1957, 54-64.

[8] L. Hörmander, *An Introduction to Complex Analysis in Several Variables*, Van Nostrand, Princeton, 1966.

[9] T. Nishino, Sur une propriété des familles de fonctions analytiques de deux variables complexes, *J. Math. Kyoto Univ.* 4 (1965), 255-282.

[10] G. Stolzenberg, *Volumes, limits and extensions of analytic varieties*, Lecture Notes in Math no 19, Springer-Verlag, Heidelberg 1966.

[11] E. L. Stout, *The Theory of Uniform Algebras*, Bogden and Quigley, Belmont, Calif., 1971.

[12] J. Wermer, *Banach Algebras and Several Complex Variables*, second edition, Springer-Verlag, New York, 1976.

University of Illinois at
Chicago Circle
Chicago

SOUS-HARMONICITÉ ET ALGÈBRES DE FONCTIONS

par

Bernard Aupetit*

Dans [13] (voir aussi [14], p. 132-140), J. Wermer a démontré le résultat suivant:

THÉORÈME 1 (Wermer). *Soient A une algèbre de Banach commutative, M son ensemble de caractères, X sa frontière de Shilov, f,g dans A alors* $\lambda \to \operatorname{Log} \max_{\chi \in \bar{f}^{1}(\lambda)} |\chi(g)|$ *est sous-harmonique sur* $\hat{f}(M)\setminus\hat{f}(X)$, *où* $\bar{f}^{1}(\lambda)$ *désigne l'ensemble des* $\chi \in M$ *tels que* $\chi(f)=\lambda$ *et* $\hat{f}$ *la transformée de Gelfand.*

$\bar{f}^{1}(\lambda)$ est appelé la *fibre* sur λ. La démonstration de théorème de Wermer utilise surtout le théorème du principe du maximum local de Rossi. Z. Słodkowski [12] en a donné une démonstration beaucoup plus élémentaire.

Si $K_g(\lambda)$ dénote l'ensemble de $\chi(g)$, pour $\chi \in \bar{f}^{1}(\lambda)$, j'ai pu montrer dans [4], en s'inspirant d'idées de [1] qui sont fondamentales pour l'étude de la structure des algèbres de Banach (voir aussi [2]), que $\lambda \to \operatorname{Log} \delta(K_g(\lambda))$, où δ désigne le diamètre, est sous-harmonique sur $\hat{f}(M)\setminus\hat{f}(X)$ d'où, toujours en reprenant les idées de [1], j'ai pu obtenir avec J. Wermer dans [4] la généralisation suivante du théorème de structure analytique de E. Bishop.

THÉORÈME 2. (Bishop-Aupetit-Wermer). *Soit A une algèbre de Banach commutative,* M *son ensemble de caractères, X sa frontière de Shilov et f dans A. Supposons que* $\hat{f}(M)\setminus\hat{f}(X)$ *est non vide et soit W une composan-*

* Travail subventionné par le Conseil national de recherches en sciences naturelles et en génie du Canada (A 7668) et le Ministère de l'Education du Québec (Subvention FCAC).

te connexe de cet ensemble. Supposons que W contient un sous-ensemble G tel que:

-1° G est de capacité extérieure strictement positive.

-2° Les fibres $f^{-1}(\lambda)$ *sont finies sur G.*

Alors il existe un entier n tel que $\# f^{-1}(\lambda) \leq n$ *pour tout* $\lambda \in W$ *et* $f^{-1}(W)$ *admet une structure analytique de variété analytique complexe de dimension 1 sur laquelle les éléments de A sont analytiques.*

La démonstration utilise la sous-harmonicité de $\lambda \to \mathrm{Log}\ \delta(K_g(\lambda))$, le théorème de Cartan de la localisation au cas n=1 qui est assez technique.

Le théorème de Bishop classique pose la condition plus forte que *G est de mesure planaire strictement positive.* Sa démonstration traditionnelle est laborieuse (voir [8] ou [14], chapitre 11, où déjà elle est sérieusement simplifiée). Ce résultat est fondamental dans le problème de l'approximation polynomiale sur les arcs non fermés C^1 ou rectifiables de $\mathbb{C}^n$ (on trouvera un exposé assez clair dans [9]).

Dans [2] et [3] nous avons fait la conjecture que $\lambda \to \mathrm{Log}\ \delta_n(K_g(\lambda))$ et $\lambda \to \mathrm{Log}\ c(K_g(\lambda))$ sont sous-harmoniques sur W où δ_n denote le n-ième diamètre et c la capacité. D. Kumagai [10] a donné une réponse partielle à cette conjecture dans le cas où l'algèbre A satisfait à la condition technique $\partial^1(A \hat{\otimes} A) = (\partial^0 A \times \partial^1 A) \cup (\partial^1 A \times \partial^0 A)$, où $A \hat{\otimes} A$ est le produit tensoriel projectif, $\partial^0 A$ le frontière de Shilov ordinaire X et $\partial^1 A$ la frontière de Shilov généralisée d'ordre 1 (voir [7] et plus loin).

En fait cette conjecture est vraie, comme Z. Słodkowski [12] et moi-même [3] l'ont montré . De ces résultats on déduit, à l'aide du théorème de Cartan, une démonstration beaucoup plus simple de théorème 2 que celle donnée dans [4].

Comme dans [4] on peut obtenir la généralisation suivante du théorème de R. Basener [6].

THÉORÈME 3 (Basener-Aupetit-Wermer). *Supposons que A, X,f,g,W vérifient les hypothèses du théorème 2 et que W contient un sous-ensemble G tel que:*

-1° G est de capacité extérieure strictement positive.

-2° les fibres $f^{-1}(\lambda)$ sont dénombrables sur G.

Alors il existe un ouvert non vide de M *admettant une structure analytique complexe de dimension 1 (c'est-même un polydisque analytique) sur lequel les éléments de A sont analytiques.*

Dans [6] , R. Basener suppose G=W auquel cas $f^{-1}(W)$ contient un ouvert dense admettant une structure analytique. Si on suppose seulement que G est de mesure planaire strictement positive, la même démonstration, comme l'a montré B. Cole, montre l'existence d'un polydisque analytique dans $f^{-1}(W)$, mais pour obtenir le théorème 3 il faut une méthode plus puissante. Dans le cas où G=W on peut donner une démonstration beaucoup plus topologique du résultat de Basener, voir [5].

En fait par des techniques encore plus poussées on peut même réussir à *globaliser* le théorème 3 pour obtenir le résultat suivant:

THÉORÈME 4(Aupetit [3]). *Supposons que A, X,f,g,W,G, vérifient les hypothèses du théorème 2. Alors $f^{-1}(W)$ contient un ouvert dense admettant une structure analytique complexe de dimension 1 sur lequel les éléments de A sont analytiques.*

Les mêmes méthodes permettent de simplifier et de généraliser le théorème de structure analytique à n-dimensions obtenu par R. Basener [7] et utilisé par N. Sibony [11] pour obtenir diverses applications.

Soit A une algèbre de fonctions dont l'espace des idéaux maximaux est M et dont la frontière de Shilov classique sera maintenant notée $\partial^0 A$. Soit $F = (f_1,\ldots,f_n)\in A^n$ et $V(F)=\{\chi\in M \mid \chi(f_1)= \ldots =\chi(f_n) = 0\}$. Par définition la *frontière de Shilov d'ordre n* est:

$$\partial^n A = \overline{\cup\, \partial^0 (A(V(F)))}$$

pour tous les $F\in A^n$ et où A(V(F)) dénote l'adhérence de l'ensemble des restrictions des éléments de A sur V(F).

THÉORÈME 5(Aupetit [3]). *Soit $F\in A^n$, W une composante connexe de $F(M)\setminus F(\partial^{n-1}A)$. Supposons que W contient un ensemble G tel que:*

-1° G n'est pas pluri-polaire, autrement dit il n'existe pas de

fonction $\phi(\lambda)$ *pluri sous-harmonique sur* $\mathbb{C}^n$ *telle que*

$G \subset \{\lambda \in \mathbb{C}^n \mid \phi(\lambda) = -\infty\}$.

-2° *les fibres* $\bar{F}^1(\lambda)$ *sont finies sur* G.

Alors il existe un entier $m \geq 1$ *tel que:*

a) $W = \bigcup_{k=1}^{m} W_k$

b) $\bigcup_{k=1}^{m-1} W_k$ *est une sous-variété analytique propre de* W.

c) $\bar{F}^1(W)$ *a une structure analytique de dimension* m *sur laquelle tout élément de* A *est analytique.*

On peut remplacer la condition 1° par la condition plus faible que G est de β-*mesure de Hausdorff* strictement positive, où $\beta > 2n-2$, ce qui évidemment permet de retrouver la condition de Basener avec $m_{2n}(G) > 0$.

Dans ce cas le théorème 4 peut se généraliser de façon semblable.

Le lecteur intéressé trouvera les détails de ces résultats dans [3], où il trouvera également de nombreuses propriétés spectrales pour les algèbres de Banach, en particulier la solution de la conjecture de Pełczyński.

BIBLIOGRAPHIE

[1] B. Aupetit, Caractérisation spectrale des algèbres de Banach de dimension finie. *J. Functional Analysis* 26 (1977), 232-250.

[2] B. Aupetit, *Propriétés spectrales des algèbres de Banach*, Lecture notes in Mathematics no 735, Springer-Verlag, Heidelberg, 1979.

[3] B. Aupetit, Subharmonicity of spectral functions, à paraître.

[4] B. Aupetit, and J. Wermer, Capacity and uniform algebras, *J. Functional Analysis* 98 (1978), 386-400.

[5] H. S. Baer and G.N. Hile, Analytic Structure in function algebras, *Houston J. Math.*, à paraître.

[6] R. Basener, A condition for analytic structure, *Proc. Amer. Math. Soc.* 36 (1972), 156-160.

[7] B. Basener, A generalized Shilov boundary and analytic structure, *Proc. Amer. Math.Soc.* 47 (1975), 98-104.

[8] E. Bishop, Holomorphic completions, analytic continuations and the interpolation of semi-norms, *Ann. Math.* 78 (1963), 468-500.

[9] P. Jakóbczak, On the existence of analytic structure in the spectrum of a uniform algebra. *Bull. Acad. Polonaise Sci., Sér. Math. Astr. Phys.*, 26 (1978).

[10] D. Kumagai , *Uniform algebras and subharmonic functions*, Ph. D.thesis, Lehigh University, 1978.

[11] N. Sibony, Multi-dimensional analytic structure in the spectrum of of a uniform algebra, in *Spaces of analytic functions Seminar*, Lecture Notes in Mathematics no 512, Springer-Verlag, Heidelberg, 1976, pp. 139-165.

[12] Z. Słodkowski, On subharmonicity of capacity. A paraître.

[13] J. Wermer, Subharmonicity and hulls, *Pacific J. Math.* 58 (1975), 283-290.

[14] J. Wermer, *Banach algebras and several complex variables*, 2nd edition. Springer-Verlag, New York, 1976.

Département de Mathématiques
Université Laval
Québec

APPROXIMATION ON PSEUDOCONVEX DOMAINS

by

Eric Bedford

and

John Erik Fornaess

Here we discuss some problems in approximation which are related to the problem of finding pseudoconvex neighborhoods. Since we omit various topics, we refer the reader to the articles of Birtel [4], Henkin and Chirka [16], and Wells [28].

If $S \subset \mathbb{C}^n$, we let $\mathcal{O}(S)$ be the functions on S which are holomorphic in a neighborhood of each point of S. All domains considered below will be of the following form:

$\bar{\Omega} = \{r<0\} \subset \mathbb{C}^n$ is bounded, r is C^∞ in a neighborhood of $\bar{\Omega}$,

$\{r=0\} = \partial\Omega$, and $dr \neq 0$ on $\partial\Omega$.

1. STRONGLY PSEUDOCONVEX DOMAINS.

The domain Ω is *pseudoconvex at the point* $z_0 \in \partial\Omega$ if

$$\sum_{i,j=1}^{n} \frac{\partial^2 r(z_0)}{\partial z_i \partial \bar{z}_j} a_i \bar{a}_j \geq 0 \qquad (1)$$

for each vector $(a_1,\ldots,a_n) \in \mathbb{C}^n$ with

$$\sum_{i=1}^{n} \frac{\partial r(z_0)}{\partial z_i} a_i = 0. \qquad (2)$$

Ω is *strongly pseudoconvex* at the point $z_0 \in \partial\Omega$ if (1) holds with strict inequality for all $a \neq 0$ satisfying (2). The domain Ω is *(strongly) pseudoconvex* if it is (strongly)pseudoconvex at all of its boundary points. By $w(\partial\Omega)$ we will denote all points of $\partial\Omega$ where Ω is not strongly pseudoconvex. If $\{z\in \mathbb{C}^n : r<0\} = \Omega$ is strongly pseudoconvex, it is immediate that

if ε is a real number sufficiently close to 0, then $\{z \in \mathbb{C}^n : r(z) < \varepsilon\}$ is strongly pseudoconvex. (3)

The ball $\{|z_1|^2 + \ldots + |z_n|^2 < 1\}$ is strongly pseudoconvex and is the local "model" for all strongly pseudoconvex points. Examples of strongly pseudoconvex domains may be obtained as follows: let $U = \{x \in \mathbb{R}^n : \rho(x) < 0\} \subset \mathbb{R}^n \subset \mathbb{C}^n$ be a smoothly bounded domain; then for k large

$$\tilde{U} = \{x+iy \in \mathbb{C}^n : \rho(x) + k|y|^2 < 0\} \subset \mathbb{C}^n$$

is a strongly pseudoconvex domain with $\tilde{U} \cap \mathbb{R}^n = U$.

It is immediate from (1) and (2) that every convex domain is pseudoconvex. A partial converse is true. Let us assume that $z_0 = 0 \in \partial\Omega$ and $\left(\frac{\partial r}{\partial z_1}(0), \ldots, \frac{\partial r}{\partial z_n}(0)\right) = (0,\ldots,0,1)$; and for each k we define new coordinates

$$\begin{cases} z_1' = z_1 \\ \quad\vdots \\ z_{n-1}' = z_{n-1} \\ z_n' = z_n + \sum_{i,j=1}^{n} z_i z_j \dfrac{\partial^2 r(0)}{\partial z_i \partial z_j} + k z_n^2 . \end{cases}$$

If the form in (1) is positive definite, then for k sufficiently large Ω is convex in the z'-coordinates. Thus

after a holomorphic change of coordinates every strongly pseudoconvex domain may be made (locally) strictly convex. (4)

The following result was obtained by Henkin [14], Kerzman [19], and Lieb [22].

THEOREM 1. *If $\Omega \subset\subset \mathbb{C}^n$ is strongly pseudoconvex, then $O(\bar{\Omega})$ is uniformly dense in $C(\bar{\Omega}) \cap O(\Omega)$.*

PROOF: We will find $f_j \in O(\bar{\Omega})$ so that

$$\lim_{j\to\infty} \max_{z\in\bar{\Omega}} |f_j(z)-f(z)| = 0.$$

There are two parts of the proof. First we approximate f by functions g_j which are smooth is a neighborhood of $\bar{\Omega}$ and "almost" holomorphic there. Then we add a small correction term to make g_j holomorphic.

We cover $\partial\Omega$ by open sets $U_1,\ldots,U_k$ such that there is a constant vector ξ_j which is transverse to $\partial\Omega$ on $\partial\Omega\cap U_j$. Let $\phi,\phi_1,\ldots,\phi_k$ be a partition of unity for a neighborhood of $\bar{\Omega}$ subordinate to the cover Ω, $U_1,\ldots,U_k$. For large j,

$$g_j(z) = \phi(z)f(z) + \sum_{\ell=1}^{n} \phi_\ell(z)f(z-\frac{\xi_\ell}{j}) \quad (5)$$

is smooth on $G_j = \{r<\varepsilon_j\}$ for some small $\varepsilon_j>0$.
Since f is continuous on $\bar{\Omega}$, g_j converges uniformly to f on $\bar{\Omega}$.

Now we note that

$$\bar{\partial}g_j = f(z)\bar{\partial}\phi(z) + \sum_{\ell=1}^{k} f(z-\frac{\xi_\ell}{j})\, \bar{\partial}\phi_\ell\,(z).$$

Since $\phi + \sum\phi_\ell = 1$ in a neighborhood of $\bar{\Omega}$, it follows that $\bar{\partial}g_j$ is uniformly small on $G_j = \{r<\varepsilon_j\}$. Finally we use the deep fact that it is possible to solve the $\bar{\partial}$ - equation with uniform estimates, i.e., there is a solution $h_j\varepsilon C^\infty(G_j)$ of $\bar{\partial}h_j = \bar{\partial}g_j$ with the property that

$$\sup_{G_j}|\le h_j| \le C \sup_{G_j} |\bar{\partial}g_j|$$

It follows, then, that $f_j = g_j - h_j \varepsilon\ \mathcal{O}(\bar{\Omega})$ and f_j converges uniformly to f on $\bar{\Omega}$. □

The key to this proof, the uniform estimates for $\bar{\partial}$, were first obtained by Grauert and Lieb [13]. The $\bar{\partial}$-equation has also been solved with estimates in the norms of L^p, $C^{m,\alpha}$ and other spaces (a survey of this is given by Henkin [15]). The proof above then gives approximation by $\mathcal{O}(\bar{\Omega})$ in the appropriate topologies, e.g. $\mathcal{O}(\bar{\Omega})$ is dense in $\mathcal{O}(\Omega)\cap L^p(\Omega)$ $1\le p<\infty$, $\mathcal{O}(\Omega)\cap C^{m,\alpha}(\bar{\Omega})$, etc. There is a slight gain of regularity in the solution of the $\bar{\partial}$-equation; it is enough that the proof above gives also bounded approximation (cf. [3]): if $f\ \varepsilon\ \mathcal{O}(\Omega)\cap L^\infty(\Omega)$, then there exist $f_j\ \varepsilon\ \mathcal{O}(\bar{\Omega})$ such that $\{f_j\}$ converges to f uniformly on compacta and such that $\|f_j\|_{L^\infty(\Omega)} \le \|f\|_{L^\infty(\Omega)}$. (This result was first obtained by Cole and Range [7] by another method.)

2. WEAKLY PSEUDOCONVEX DOMAINS: EXAMPLES. We want to discuss the possibility of dropping the hypothesis of strong pseudoconvexity from Theorem 1, so we give some examples of weakly pseudoconvex domains. In particular, we show that the analogues of (3) and (4) do not hold in the weakly pseudoconvex case.

First we consider the bounded domain $\Omega \subset \mathbb{C}^2$ given by

$$\Omega = \{(z,w)\in\mathbb{C}^2 \colon \operatorname{Re} w + |z|^6 + t\operatorname{Re}\ z\bar{z}^5 + |z|^8 + |w|^8 < 0\} .$$

It is easily seen that Ω is pseudoconvex it $|t| \leq 9/5$. On the other hand, Ω is not convex, if $|t| > 1$. This gives an examples of the phenomenon first noted by Kohn and Nirenberg [21]:

there is no holomorphic change of coordinates which makes Ω convex at $(0,0)$.

If there were, there would have to be a germ of a complex manifold $M = \{w = f(z)\}$ such that $M \cap \bar{\Omega} = (0,0)$. But Ω is invariant under the transformation $(z,w) \longrightarrow (iz,w)$ and so we may take $f(z)$ to have the form

$$f(z) = a_0 + a_4 z^4 + a_8 z^8 + \dots$$

By checking the order of contact of M to $\partial\Omega$ at $(0,0)$, it follows that $a_0 = a_4 = 0$. But since Ω is nonconvex and tangent to $\{w=0\}$ to order 6 at $(0,0)$, it follows that $M \cap \Omega \neq \emptyset$. Thus Ω cannot be made convex at $(0,0)$.

It follows that the existence of local peak functions at $(0,0)$ is not trivial, although they have been shown to exist (see [2]).

Now we show that (3) fails in the weakly pseudoconvex case. The following smoothly bounded pseudoconvex domain in $\mathbb{C}^2$ was introduced by Diederich and Fornaess [10]:

$$W(k,c) = \{(z,w)\in\mathbb{C}^2 : |w-e^{ic \log|z|}|^2 - 1 + \lambda(|z|) < 0\} ,$$

where $\lambda(t)$ is a sufficiently convex, smooth function on $(0,\infty)$ such that $\lambda^{-1}(0) = [e^{-k}, e^{k}]$ and $\lim_{t\to 0,\infty} \lambda(t) = \infty$.

As was shown in [10], $w(\partial W(k,c))$ is precisely the annulus

$$A = \{(z,0);\ -k < \log|z| < k\}.$$

It was also shown in [10] that

> *if $2kc > \pi$, then there is an open set $W_1 \supset A \cup W$, $W_1 \neq W$ such that every $f \in O(\bar{W})$ has an analytic extension to W_1.* (7)

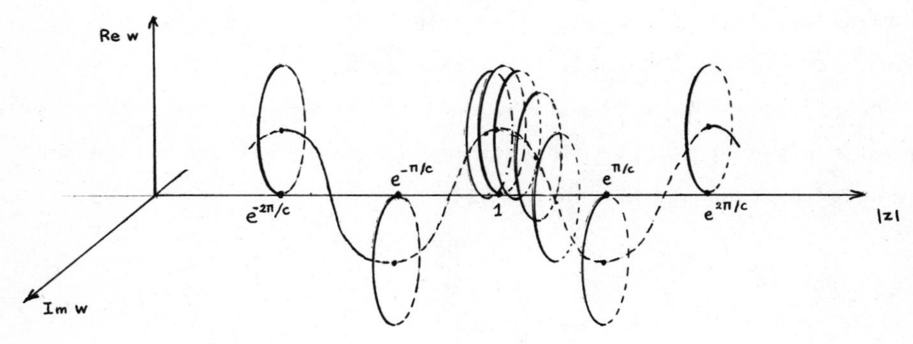

For $t \in \mathbb{C}$ in some small neighborhood of $(-\varepsilon,1]$, let $A(t)$ be the annulus

$$A(t) = \{(z,t) \; ; \; -\pi/2c<\log|z|<\pi/2c\}.$$

It $f \in O(\bar{W})$, we may extend f to $\cup A(t)$ by the Cauchy integral formula

$$f(z,t) = \frac{1}{2\pi i}\int_{\partial A(t)} \frac{f(\zeta,t)}{\zeta - z}\, d\zeta \quad \text{since } A(0)\cup\bigcup \partial A(t)\subset W(k,c).$$

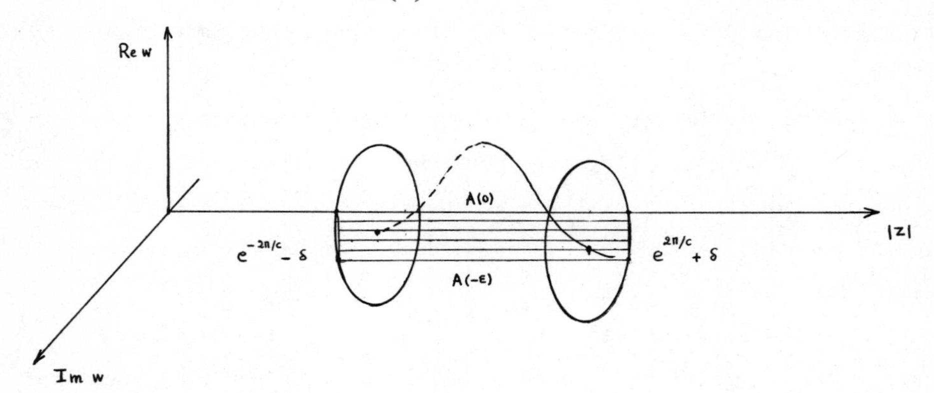

It follows that two types of approximation fail for $W(c,k)$ when $2ck > \pi$. First, $O(\bar{W})$ is not dense in $O(W)\cap C(\bar{W})$. Further, $O(\bar{W})$ is not dense in $\mathcal{O}(W)$, with the topology of uniform convergence on compact subsets of W. In fact, W in not Runge in W_1. To see this, one may show that there is a set of the form $K = (A_1 \cup A_2) \times \{1/2\} \subset W$, where A_1 and A_2 are compact annuli in the z-plane centered at the origin, and such that K is holomorphically convex in w but not in w_1. By (7), then, there is no domain $W_2 \supset \supset W$ such that W is Runge in W_2.

Now let us consider the more general situation where $\Omega \subset \mathbb{C}^2$ is a smoothly bounded pseudoconvex domain, and there exists a holomorphic embedding of a smoothly bounded open set $E' \subset \mathbb{C}$, $\phi : E' \longrightarrow E \subset \partial\Omega$. Let us assume ϕ extends to a diffeomorphism between $\bar{E}'$ and $\bar{E}$, and that $\bar{E} = w(\partial\Omega)$. Then (see [3]) there exists a C^3 function f such that $\bar{\partial} f(z) = O(\mathrm{dist}(z,\bar{E})^2)$ and $df \neq 0$ on $\bar{E}$.

It follows that ∂f is a multiple of ∂r, where r is a defining function of Ω, i.e.

$$\partial r = e^{i\theta+\mu}\partial f, \tag{8}$$

where θ, $\mu \in C^2(\bar{E})$ are real functions. The geometric significance of θ is that it measures the winding of the holomorphic normal vector ∂f of E with respect to the normal vector to $\partial\Omega$. It is important to note that since Ω is pseudoconvex one may show that θ is harmonic.

In case E is doubly connected, we may make a holomorphic change of coordinates in a neighborhood of E so that

$$E = \{(z,\omega)\ ;\ -\kappa < \log|z| < \kappa, \omega = 0\}$$

Furthermore, replacing w by $we^{g(z)}$, we may replace $\theta(z)$ by $\theta - \mathrm{Im}\, g = \gamma \log|z|$. If $\kappa < k$ and $\gamma = c$, then it is possible to fit the germ of $W(k,c)$ at E inside of Ω with first order contact along E. It follows that (7) also holds for Ω. We note that κ is the modulus of E, and

$$\gamma = -\frac{1}{\pi}\mathrm{Im}\int_{|z|=1} \bar{\partial}\theta$$

mesures the "rates" at which $\partial\Omega$ winds around E.

THEOREM 2. *If Ω,E are as above, and if $\gamma\kappa > \pi/2$, then (7) holds for Ω, and $\mathcal{O}(\bar{\Omega})$ is not dense in $\mathcal{O}(\Omega) \cap C(\bar{\Omega})$.*

Idea of proof: By the preceding discussion, we may embed a germ of a domain $W(c,k)$ inside Ω along **E**. If $2ck > \pi$, then $W(c,k)$ has the desired properties, and so does Ω. □

3. WEAKLY PSEUDOCONVEX DOMAINS: APPROXIMATION. We will discuss some instances in which the proof of Theorem 1 may be extended to the weakly pseudoconvex case. It was observed by Range [25] that the proof may be refined to work in the case where $w(\partial\Omega)$ consists of isolated points. We say that the vector field

$$v = (f_1, g_1, \ldots, f_n, g_n) = (f_1 + ig_1, \ldots, f_n + ig_n)$$

is *holomorphic* if $f_j + ig_j$ is holomorphic for $1 \le j \le n$. Fornaess and Nagel [12] established the following.

THEOREM 3. *If $\Omega \subset\subset \mathbb{C}^n$ is a smoothly bounded pseudoconvex domain, and if there exists a holomorphic vector field in a neighborhood of $w(\partial\Omega)$, which is transverse to $\partial\Omega$, then $O(\bar{\Omega})$ is dense in $O(\Omega) \cap C(\bar{\Omega})$. Further, it follows that*

$$\text{there are pseudoconvex domains } \bar{\Omega} \subset \Omega_j \subset \mathbb{C}^n \text{ such that } \bigcap_{j=1}^{\infty} \Omega_j = \bar{\Omega} \tag{9}$$

It is immediate that in condition (9) the Ω_j may be taken to be strongly pseudoconvex with smooth boundary.

COROLLARY. *If $\Omega \subset\subset \mathbb{C}^2$ has real analytic boundary, then $O(\bar{\Omega})$ is dense in $O(\Omega) \cap C(\bar{\Omega})$.*

EXAMPLE: Consider the domain

$$\Omega = \{(z, w, \eta) \in \mathbb{C}^3 : |z|^4 + |w|^2 + |\eta|^2 + |\tfrac{1}{\eta}|^2 < 4\}$$

It may be shown that $w(\partial\Omega) = \{z=0\} \cap \partial\Omega$ and that there is no tranverse holomorphic vector field in a neighborhood of $w(\partial\Omega)$. On the other hand, $O(\bar{\Omega})$ is dense in $O(\Omega) \cap C(\bar{\Omega})$, as may be seen by expanding $f(z,w,\eta)$ in Laurent series.

We will say that Ω is uniformly H-convex if Ω satisfies (3) and if there is a constant $C > 0$ such that

$$\frac{1}{Cj} \le \operatorname{dist}(\bar\Omega, \partial\Omega_j) \le \sup_{z\in\partial\Omega_j} \operatorname{dist}(\Omega, z) \le \frac{C}{j} . \qquad (10)$$

(see Weinstock [27] .)

The following result, due to Čirka [6] , is similar in spirit to Theorem 1, except that L^2-estimates for the $\bar\partial$-equation are used instead of L^∞- estimates. This method has also been used by Hörmander-Wermer [18], Nirenberg-Wells [23], and others.

THEOREM 4. *Let $\Omega \subset\subset \mathbb{C}^n$ be uniformly H-convex. If $f \in O(\Omega) \cap C^{n+1}(\bar\Omega)$, then there are $f_j \in O(\bar\Omega)$ such that*

$$\lim_{j\to\infty} \| f - f_j \|_{L^\infty(\Omega)} = 0.$$

PROOF: We may assume that f is extended to a C^{n+1} function $\tilde f$ on a neighborhood of $\bar\Omega$, so

$$|\bar\partial \tilde f(z)| \le C(\operatorname{dist}(z, \bar\Omega))^n .$$

By the L^2- estimates of Hörmander [17], there exists $h_j \in L^2(\Omega_j)$ such that $\bar\partial h_j = \bar\partial \tilde f$ on Ω_j and such that

$$\| h_j \|_{L^2(\Omega_j)} \le C \| \bar\partial \tilde f \|_{L^2(\Omega_j)}$$

and such by (10),

$$\| h_j \|_{L^2(\Omega)} = 0(j^{-n-1/2})$$

where the constant C depends only on the diameter of Ω_j.

Clearly $f_j = \tilde f - h_j \in O(\Omega_j)$. It follows that f_j approximates f in $L^\infty(\Omega)$ by the following standard elliptic estimate

$$\| h_j \|_{L^\infty(\Omega)} \le C_2 [\operatorname{dist}(\bar\Omega, \partial\Omega_j))^{-n} \| h_j \|_{L^2(\Omega_j)} + \operatorname{dist}(\bar\Omega, \partial\Omega_j) \| \bar\partial h_j \|_{L^\infty(\Omega_j)}]$$

which gives, by (10).

$$\| h_j \|_{L^\infty(\Omega)} = 0(j^{-\frac{1}{2}}) . \ \square$$

Thus it is quite reasonable that the following has arisen:

CONJECTURE 1. *If $\Omega \subset\subset \mathbb{C}^n$ is a smoothly bounded pseudoconvex domain, then $O(\bar\Omega)$ is dense in $O(\Omega) \cap C(\bar\Omega)$ if and only if (9) holds.*

A condition like (9) is used for Runge approximation. If there is a 1-parameter family of pseudoconvex domains $\{\Omega_t\}$, $0 \le t \le 1$, then Ω_0 is Runge in Ω_1 (see Docquier-Grauert [8] for a precise statement). In connection with (9), it is natural to define the Nebenhülle $N(\Omega) = \cap \Omega_\alpha$, where the intersection is taken over all pseudoconvex domains $\bar{\Omega} \subset \Omega_\alpha \subset \mathbb{C}^n$. We also define $H(\Omega)$ as the $C(\bar{\Omega})$-closure of $O(\bar{\Omega})$. One may show that *each* $f \in H(\Omega)$ *has an analytic continuation to the interior of* $N(\Omega)$.

Catlin [5] has shown that if $\Omega \Subset \mathbb{C}^n$ is smoothly bounded and pseudoconvex, then there is a function $f \in C^\infty(\bar{\Omega}) \cap O(\Omega)$ which has no holomorphic continuation over any $z_0 \in \partial\Omega$. It follows, then, that

if $N(\Omega)$ contains an open subset
of $\partial\Omega$, then $H(\Omega) \subsetneq C(\bar{\Omega}) \cap O(\Omega)$.

It was shown in [9] and [11] that (9) always holds if Ω has real analytic boundary. Thus we arrive at

CONJECTURE 1'. *If* $\Omega \subset\subset \mathbb{C}^n$ *has smooth, real analytic, pseudoconvex boundary, then* $H(\Omega) = O(\Omega) \cap C(\bar{\Omega})$.

We remark that half of Conjecture 1 is an obvious consequence of

CONJECTURE 2. *Let* $\mathbb{C}^n \supset\supset \Omega_1 \supset \Omega_2 \supset \dots$ *be strongly pseudoconvex domains with* $\cap \Omega_j = \bar{\Omega}$. *If* g *is a (0,1)-form with* $\bar{\partial} g = 0$ *on* Ω_j, *then there exists* f *such that* $\bar{\partial} f = g$ *on* Ω_j *and*

$$\|f\|_{L^\infty(\Omega_j)} \le C \|g\|_{L^\infty(\Omega_j)} .$$

(C is independent of j.)

The following is also well-known.

CONJECTURE 2'. *If* $\Omega \subset\subset \mathbb{C}^n$ *is smoothly bounded and pseudoconvex, and if* g *is a (0,1)-form with bounded coefficients such that* $\bar{\partial} g = 0$, *then there exists* $f \in L^\infty(\Omega)$ *such that* $\bar{\partial} f = g$.

A reasonable approach to Conjecture 2' is to consider the case where $\partial\Omega$ is real analytic. Some results have been obtained in this direction, see Range [26] and Kohn [20].

4. APPROXIMATION ON GENERAL PSEUDOCONVEX DOMAINS. Since in général $O(\bar{\Omega})$ is not dense in $C(\bar{\Omega}) \cap O(\Omega)$ the following two results are of

interest. First we let $H^j(\Omega)$ denote the Sobolev space with norm

$$\|f\|^2_{H^j(\Omega)} = \sum_{|\alpha|\le j} \|D^\alpha f\|^2_{L^2(\Omega)} .$$

THEOREM 5 (Catlin [5]). *If $\Omega \subset\subset \mathbb{C}^n$ is smoothly bounded and pseudoconvex, then $C^\infty(\bar{\Omega}) \cap O(\Omega)$ is dense in $H^j(\Omega) \cap O(\Omega)$.*

Next is a result on uniform approximation (cf. Range [24]).

THEOREM 6 (Beatrous [1]). *If $\Omega \subset\subset \mathbb{C}^n$ is smoothly bounded and pseudoconvex, then $O(\bar{\Omega}\backslash w(\partial\Omega)) \cap C(\bar{\Omega})$ is dense in $O(\Omega) \cap C(\bar{\Omega})$.*

5. EXAMPLES WHERE APPROXIMATION HOLDS. Let us discuss some cases where it is possible to find a holomorphic transverse vector field in a neighborhood of $w(\partial\Omega)$. For simplicity, we consider $\Omega \subset \mathbb{C}^2$; a more general treatment is given in [3]. We will first consider the case where $w(\partial\Omega)$ is the closure of a 1-dimensional complex manifold M.

The main point of this section is that the possibility of approximation depends on the conformal invariants of M and the winding of $\partial\Omega$ around M. We express this in terms of the cohomology class of Im $\bar{\partial}\theta$ in $H^1(M,\mathbb{R})$ and the moduli of M. First we consider the topologically trivial case, i.e. M is a disk.

THEOREM 7. *Let $\Omega \subset\subset \mathbb{C}^2$ be a smoothly bounded pseudoconvex domain, and let there be a smooth embedding $\phi : \{|\zeta|\le 1\} \longrightarrow \partial\Omega$ such that $\phi(|\zeta|\le 1) = w(\partial\Omega)$, and ϕ is holomorphic on $\{|\zeta| < 1\}$. Then (9) holds, and $O(\bar{\Omega})$ is dense in $O(\Omega) \cap C(\bar{\Omega})$.*

Idea of proof: Recall f,r and θ from the end of §2. Since θ is harmonic, and has a harmonic conjugate θ^*, it follows that we may replace ∂r by a holomorphic multiple of ∂f. We now approximate ∇r by a holomorphic vector field in a neighborhood of $w(\partial\Omega)$. The result now follows from Theorem 2. □

Now we consider the problem of a general Riemann surface M. We want to repeat the procedure of finding a harmonic conjugate of θ, but this in general will not exist. Therefore, we consider the following minimization:

$$\mu = \inf \; \|\theta + \mathrm{Re} h\|_{L^\infty(M)} \tag{11}$$

where the infimum is taken over

$h \in \mathcal{O}(M) \cap C(\bar{M})$.

In case M is an annulus, we may assume $M = \{\zeta \in \mathbb{C} ; -\kappa < \log|\zeta| < \kappa\}$. It is easily seen that the function assuming the infimum in (11) is

$$\theta + \mathrm{Re} h = \frac{\mu}{\kappa} \log|\zeta| .$$

It follows that

$$\mu = -\frac{\kappa}{\pi} \mathrm{Im} \int_{|\zeta| = 1} \bar{\partial}\theta \; ;$$

and so we have a complement to Theorem 2.

THEOREM 8. *If $\mu < \frac{\pi}{2}$, then $\bar{\Omega}$ has a neighborhood system of pseudoconvex open sets. Moreover, $\mathcal{O}(\bar{\Omega})$ is dense in $\mathcal{O}(\Omega) \cap C(\bar{\Omega})$.*

Remark: For the converse, one would like to show that if $\mu \geq \frac{\pi}{2}$, then there is a sufficiently "large" domain $\mathcal{W}(k,c)$ inside Ω. To do this we consider the problem:

> Find $h \in \mathcal{O}(\bar{M}) \cap C(\bar{M})$ and a partition of ∂M into disjoint closed sets C_0 and C_1 such that $\theta + \mathrm{Re} h \geq \frac{\pi}{2}$ on C_1 and $\theta + \mathrm{Re} h \leq -\frac{\pi}{2}$ on C_0. (12)

There is one further situation we consider. If $w(\partial\Omega)$ contains an open subset of $\partial\Omega$, then this is foliated by Riemann surfaces, i.e. $\mathrm{int}(w(\partial\Omega)) = \bigcup_{M \in \mathcal{M}} M$.

We will need to assume that the leaves of $\mathcal{M}$ are closed.

> There exists a function $f(t,z) \in C([0,1] \times w(\partial\Omega))$ which is holomorphic in z in a neighborhood of $w(\partial\Omega)$ such that $\mathrm{int}(w(\partial\Omega)) = \bigcup_{0<t<1} M_t$, where $M_t = \{z \in \mathrm{int}(w(\partial\Omega)) ; f(t,z) = 0\}$. (13)

THEOREM 9. *If $w(\partial\Omega) = \overline{\text{int } w(\partial\Omega)}$ satisfies (13), and if $w(\partial\Omega)$ has a neighborhood basis of pseudoconvex sets $\{w_j\}_{j=1}^{\infty}$ such that for some constant c, $\{z: dist(z,w(\partial\Omega)) < \frac{1}{cj}\} \subset w_j \subset \{z: dist(z,w(\partial\Omega)) < \frac{c}{j}\}$, then $\bar{\Omega}$ has a neighborhood basis of pseudoconvex sets and $\mathcal{O}(\bar{\Omega})$ is dense in $\mathcal{O}(\Omega) \cap C(\bar{\Omega})$.*

If the leaves are not assumed to be closed, approximation may fail: Let $S_R = \{(z,w) \in \mathbb{C}^2;\ 1 \leq |\zeta| \leq R\ , |w| \leq 1$ and $\mathrm{Re}(we^{i\log z\bar{z}}) = 0\}$, which is a Levi flat hypersurface in $\mathbb{C}^2$. One can find a pseudoconvex domain Ω such that $S_R = w(\partial\Omega)$. If $R \geq e^{\pi}$ one can show (see [3]) that Ω has a strictly larger nebenhuelle and $\mathcal{O}(\bar{\Omega})$ fails to be dense in $\mathcal{O}(\Omega) \cap C(\bar{\Omega})$. In this example the leaves M_t are described by

$$M_t = \{w = it\ z^{-2i} : 1 < |z| < R\ ,\ |w| < 1\}$$

REFERENCES

[1] F. Beatrous, Boundary value algebras, *Dissertation*, Tulane University, 1978.

[2] E. Bedford and J. E. Fornaess, A construction of peak functions on weakly pseudoconvex domains, *Ann. of Math.* 107 (1978), 555-568.

[3] E. Bedford and J.E. Fornaess, Domains with pseudoconvex neighborhood systems, *Inventiones Math.* 47 (1978), 1-27.

[4] E.T. Birtel, Holomorphic approximation to boundary value algebras, *Bull. A.M.S.* 84 (1978), 406-416.

[5] D. Catlin, Boundary behavior of holomorphic functions on weakly pseudoconvex domains, *Thesis*, Princeton University, 1978.

[6] E. Cirka, Approximation of holomorphic functions on smooth manifolds in $\mathbb{C}^n$, *Math, USSSR-Sb.* 7 (1969), 95-113.

[7] B. Cole and M. Range, A-measures on complex manifolds and some applications, *J. Functions Analysis* 11/4 (1972), 393-400.

[8] F. Docquier and H. Grauert, Levisches Problem und Rungescher Satz für Teilgebiete Steinscher Mannigfaltigkeiten, *Math. Ann.*

140 (1960), 94-123.

[9] K. Diederich and J. Fornaess, Pseudoconvex domains: existence of Stein neighborhood, *Duke Math.* J. 44 (1972), 641-662.

[10] K. Diederich and J. Fornaess, Pseudoconvex domains: an example with nontrivial Nebenhülle, *Math. Ann.* 225 (1977), 275-292.

[11] K. Diederich and J. Fornaess, Pseudoconvex domains with real analytic boundaries, *Ann. of Math.* 107 (1978), 371-384.

[12] J. Fornaess and A. Nagel, The Mergelyan property for weakly pseudo-convex domains, *Manuscripta Math.* 22 (1977), 199-208.

[13] H. Grauert and K. Lieb, Das Ramirersche Integral und die Lösung der Gleichung $\bar{\partial}f = \alpha$ im Bereich der beschränkten Formen, *Rice University Stud.* 56, No. 2 (1970), 29-50.

[14] G. Henkin, Integral representations of functions holomorphic in strictly pseudoconvex domains and some applications, *Math. USSSR-Sb.* 7,4 (1969), 597-616.

[15] G. Henkin, H. Lewy's equation and analysis on a pseudoconvex manifold I., *Uspehi Mat. Nauk* 32 (1977), 57-117 (= *Russ. Math. Surveys* 32 (1977).

[16] G. Henkin and E.M. Cirka, Boundary properties of holomorphic functions of several complex variables, *J. Soviet Math.* 5 (1976), 612-679.

[17] L. Hörmander, L^2-estimates and existence theorems for the $\bar{\partial}$-operator, *Acta Math.*, 113, (1965), 89-152.

[18] L. Hörmander and J. Wermer, Uniform approximation on compact sets in $\mathbb{C}^n$, *Math. Scand.*, 23, No. 1 (1968), 5-21.

[19] N. Kerzman, Hölder and L^p-estimates for solutions of $\bar{\partial}u = f$ in strongly pseudoconvex domains, *Comm. Pure. Appl. Math.* 24 (1971) 301-379.

[20] J. Kohn, Subellipticity of the $\bar{\partial}$-Neumann problem on pseudo-convex domains: sufficient conditions, *Acta Math.*

[21] J. Kohn and L. Nirenberg, A pseudoconvex domain not admitting a holomorphic support function, *Math. Ann.*, 201, No. 3 (1973), 265-268.

[22] I. Lieb, Die Cauchy-Riemannschen Diffrentialgleichungen auf streng pseudokonvexen Gebieten, *Math. Ann.* 190 (1970), 6-45

[23] R. Nirenberg and R.O. Wells, Jr., Approximation theory on differentiable submanifolds of a complex manifold, *Trans. Amer. Math. Soc.* 142 (1969), 15-35.

[24] R.M. Range, Holomorphic approximation near strictly pseudoconvex boundary points, *Math. Ann.* 201 (1973), 9-17.

[25] R.M. Range, Approximation by holomorphic functions on pseudoconvex domains with isolated degeneracies, *Manuscripta Math.*

[26] R. M. Range, On Hölder estimates for $\bar{\partial}u = f$ on weakly pseudoconvex domains.

[27] B. Weinstock, Some conditions for uniform H-convexity, *Illinois J. Math.* 19 (1975), 400-404.

[28] R.O. Wells, Jr., Function theory on diffrentiable submanifolds, in: Contributions to Analysis. *A collection of Papers Dedicated to Lipman Bers*, Academic Press, Inc., New York (1974).

Departement of Mathematics
Princeton University

POINT EVALUATIONS, APPROXIMATION IN THE MEAN AND ANALYTIC CONTINUATION

by

James E. Brennan

1. INTRODUCTION

Let Ω be a bounded simply connected domain in the complex plane $\mathbb{C}$, let dA denote two-dimensional Lebesgue measure and let w be a bounded mesurable function defined on Ω. We shall assume that $w \geq 0$ and for each p, $1 \leq p < \infty$, we shall consider the following two spaces:

(i) $H^p(\Omega, wdA)$, the closure of the polynomials in $L^p(wdA)$;

(ii) $L^p_a(\Omega, wdA)$, the set of functions in $L^p(wdA)$ which are analytic in Ω.

For a large class of measures (cf. [5, p.175]) it is known that $H^p(\Omega, wdA) \subset L^p_a(\Omega, wdA)$ and it is an old problem to determine which of these has the added feature that $H^p(\Omega, wdA) = L^p_a(\Omega, wdA)$. Whenever this happens the polynomials are said to be *complete* in $L^p_a(\Omega, wdA)$. For simplicity we shall assume that w is strictly positive and continuous on Ω so that in every case we consider the inclusion $H^p(\Omega, wdA) \subset L^p_a(\Omega, wdA)$ is trivial.

Generally speaking, the functions is $H^p(\Omega, wdA)$ are analytic in Ω because, on every compact subset of Ω, an inequality of the form

$$|Q(\xi)| \leq K \|Q\|_{L^p(wdA)} \tag{1.1}$$

is satisfied with a fixed constant K for all polynomials Q. Thus, if a sequence of polynomials converges in the $L^p(wdA)$ norm the convergence is uniform on every compact subset Ω and hence the limit function is analytic. Whenever the inequality (1.1) is satisfied at some point ξ for every polynomial Q we shall say that $H^p(\Omega, wdA)$ has a *bounded point evaluation* at ξ. It may or may not happen that $H^p(\Omega, wdA)$ has bounded point evaluations at some points in $\partial\Omega$. Attention was first drawn to this fact by Mergeljan [27] in 1955 and since then the phenomenon has been studied in greater detail by Sinanjan [34],[35], [25] and the author [5], [6] , [8] , [9].

The purpose of this survey is to acquaint the reader with some ideas and recent results in this area and to indicate the fundamental connection between the existence or non-existence of bounded point evaluations on the boundary and the completeness of the polynomials.

2. APPROXIMATION ON CERTAIN NON-CARATHÉODORY DOMAINS

Completeness questions have a long history dating back to the late 1800's and the works of Runge and Weierstrass. However, the first results specifically related to the question which is of interest to us were obtained by Carleman [10] in 1923. He proved that if Ω is a Jordan domain then $H^p(\Omega, dA) = L^p_a(\Omega, dA)$. Ten years later Markuševič and Farrell (cf. [14] , [24] and [26, p.112]) obtained the corresponding theorem for Carathéodory domains and Sinanjan [33] subsquently extended it to closed Carathéodory sets. We recall that a Carathéodory domain (or set) is a domain (or set) whose boundary coincides with the boundary of the unbounded complementary component of its closure. Thus, by 1934 it was thought that the completeness question for $H^p(\Omega, dA)$ could perhaps be answered in a purely topological way.

It was difficult to imagine how completeness could occur within a class of regions for which the following are typical examples:

(A) a Jordan domain with a cut or incision in the form of a simple arc form an interior point to a boundary point;

(B) a crescent or region topologically equivalent to one bounded by two internally tangent circles.

In the first case completeness obviously fails, at least if the cut has two-dimensional measure zero. However, Keldyš [22] (cf. also [26,p.116]) soon discovered that for a crescent Ω the polynomials may of may not be complete in $L_a^p(\Omega,dA)$ depending on the thickness of the region near the multiple boundary point. This took place around 1939 and there the matter stood for nearly a decade. Not until 1947-48 and with severe restrictions on $\partial\Omega$ was a condition found that is both necessary and sufficient for completeness in this setting. That was due to the combined efforts of M.M. Džrbašjan [13], who established sufficiency and A.L. Šaginjan [29], who established necessity (cf. also [26, p.158]). No essentially new results were obtained for another twenty years until the subject was once again taken up in earnest, this time by Havin and Maz'ja [17],[18] and by the author [5],[6]. The following results of the author are indicative of the situation as it now stands.

A. THE CRESCENT. We shall suppose that Ω is a crescent formed by two internally "tangent" Jordan curves and that U and Ω_∞ are the bounded and unbounded complementary components of $\bar{\Omega}$, respectively. For each $z\in\mathbb{C}$ we shall denote by $\delta(z)$ the Euclidean distance from z to Ω_∞.

THEOREM 2.1 . *If ∂U is a C^1 curve whose exterior unit normal $n(z)$ satisfies a Lipschitz condition $|n(z_1)-n(z_2)|\le c|z_1-z_2|$ for all z_1, $z_2\in\partial U$ then $H^p(\Omega,dA)=L_a^p(\Omega,dA)$ for every p if and only if*

$$\int_{\partial U} \log\,\delta(z)\,|dz| = -\infty\ . \tag{2.1}$$

It is interesting to note that (2.1) is independent of p and so if $H^1(\Omega,dA)=L_a^1(\Omega,dA)$ it follows that $H^p(\Omega,dA)=L_a^p(\Omega,dA)$ for all p. It is not known if this is also the case in the absence of smoothness restrictions on ∂U. In that generality, however, the following is known (cf. [7]).

THEOREM 2.2. *Let $x_o\in U$ and let $d\mu$ be the harmonic measure on ∂U representing the point x_o. There exists a fixed constant $\tau>0$ such that if*

$$\int_{\partial U} \log\,\delta(z)\,d\mu(z)=-\infty$$

then $H^p(\Omega,dA)=L^p_a(\Omega,dA)$ *whenever* $p<3+\tau$.

The proof of Theorem 2.2 proceeds generally along the same lines as the sufficiency part of Theorem 2.1 but depends on some deep results in the theory of conformal mapping. We shall be content here to outline only the proof of the first result.

Proof of Theorem 2.1. In order to establish sufficiency we first assume that condition (2.1) is satisfied and we fix a point $x_o \in U$. Because ∂U is "smooth," the harmonic measure μ for x_o on ∂U is boundedly equivalent to arc length $|dz|$. Hence $\int \log\delta(z)d\mu(z)=-\infty$. We now suppose that k is any function in $L^q(\Omega,dA)$, $q=p/(p-1)$, with the property that $\int Qk\, dA = 0$ for every polynomial Q. This implies that the Cauchy transform

$$\hat{k}(\xi)=\int \frac{k(z)}{z-\xi}\,dA\equiv 0$$

in Ω_∞ and, since $\bar{\Omega}$ has only two complementary components, it follows from a well known theorem (cf. [26, p.114]) that we need only verify that $\hat{k}\equiv 0$ in U. For this we let ϕ be a conformal map of U onto the open unit disk with $\phi(x_o)=0$. We choose a sequence $r_j \uparrow 1$ and we let U_j be the region bounded by the level set $|\phi| = r_j$. The proof that $\hat{k}\equiv 0$ in U is based on the identity

$$\int_{\partial U_j} \log|\hat{k}(z)|d\mu_j = \varepsilon \int_{\partial U_j} \log\delta(z)d\mu_j + \int_{\partial U_j} \log\left(\frac{|\hat{k}(z)|}{\delta(z)^\varepsilon}\right)d\mu_j,$$

where $d\mu_j$ is the harmonic measure for x_o on ∂U_j. By assumption the first integral on the right approaches $-\infty$ as $j\to\infty$ and it can be shown that the second integral is bounded above for a suitable choice of $\varepsilon>0$. Since $\log|\hat{k}|$ is subharmonic in U it follows that $\hat{k}\equiv 0$ there.

The proof of necessity is based on a old idea of Šaginjan (cf.[26, p.121]). We first construct a Jordan curve γ lying in Ω, having the same smoothness as ∂U and surrounding the bounded complementary component U. This is done in such a way that there is a fixed constant $\rho>0$ so that for every $z\in\gamma$ the disk with center at z and radius $\rho\delta^4(z)$ is contained in Ω (cf.[5, p.183]). Thus, by the area mean value theorem,

$$|Q(z)| \leq \frac{C}{\delta(z)^{8/p}} \|Q\|_{L^p(\Omega,dA)}$$

for every polynomial Q and all $z\epsilon\gamma$. Assuming that $\int_{\partial U}\log\delta(z)|dz|>-\infty$ it can be shown that $\int_\gamma \log\delta(z)|dz|>-\infty$ and hence there is a function h analytic and nowhere zero inside γ with the property that $|h(z)|=\delta(z)^{8/p}$ on γ. Thus, if $\{Q_j\}_{j=1}^\infty$ is a sequence of polynomials which converges in the $L^p(\Omega,dA)$ norm then $\sup_\gamma |hQ_j|\leq K$ for some fixed constant K and $j=1,2,\ldots$. Consequently, $\{hQ_j\}_{j=1}^\infty$ is a uniformly bounded sequence of analytic functions inside γ and so some subsequence converges uniformly on compact subsets of the region to an analytic function. It follows that the limit of the Q_j's is also analytic inside γ and therefore $H^p(\Omega,dA)\neq L_a^p(\Omega,dA)$. □

The proof of the preceding theorem illustrates a point that we wish to emphasize, namely: If $H^p(\Omega,dA)\neq L_a^p(\Omega,dA)$ then every $f\epsilon H^p(\Omega,dA)$ extends analytically across the boundary of Ω. This is a very general phenomenon and has been observed in a wide variety of special cases by Keldyš [22],[23], Šaginjan [29], Mergeljan [26], Mergeljan and Tamadjan [28], Shapiro [32], Havin [16], Beurling [3], Sinanjan [35] and the author [5], [6]. Additional information can also be found in the survey article [26]. Before discussing the continuation property in the most general situation we shall say a few words about another of the special cases mentioned at the outset.

B. DOMAINS WITH CUTS. Let us suppose that Ω is a domain with a single rectifiable cut γ. Thus $H^p(\Omega,dA)\neq L_a^p(\Omega,dA)$ for any p and in order to talk about completeness we must introduce a weighted measure wdA. Roughly speaking, $H^p(\Omega,wdA)=L_a^p(\Omega,wdA)$ if $w(z)\to 0$ sufficiently rapidly at every point of γ. However, this is not the only factor that must be taken into consideration. It is not true, for example, that $H^p(\Omega,wdA)=$ $=L_a^p(\Omega,wdA)$ whenever Ω is a Carathéodory domain, even if w is bounded. In general, it is necessary to impose additional restrictions on the weight w (cf. [26 , p.134] and [31, p.334]).

Here we shall follow established custom; we shall assume that $w \in L^{\infty}$ and we shall require that if ϕ maps Ω conformally onto the open unit disk D then

$$H^p(D,w(\psi)dA)=L_a^p(D,w(\psi)dA) \qquad (A_p)$$

where $\psi=\phi^{-1}$. This, in effect, makes the problem conformally invariant and it is easy to see that if $\phi^n(\phi')^{2/p} \in H^p(\Omega,wdA)$ for $n=0,1,2,...$ then $H^p(\Omega,wdA)=L_a^p(\Omega,wdA)$ (cf. [19 , p.121] and [26, p.136]). Unfortunately, it is difficult to tell in any given instance if property (A_p) is satisfied and our results are therefore limited in this regard. We remark only that w has property (A_p) for every p if $w(\psi)$ is constant on every circle $|z|=r(1-\varepsilon<r<1)$; that is, if w depends only on Green's function near $\partial\Omega$. This fact can be used in conjunction with Theorem 2.3 below to construct an abundance of weights which have property (A_p) and which decay fast enough at the cut or "inner boundary" so that completeness occurs (cf. [23, p.14] and [6, p.136]).

The following two results can be found in [6] and they more or less summarize the situation as it now stands. Results similar to Theorem 2.3 were obtained earlier by Keldyš [23],[26 , p.136], Džrbašjan [26, p.144] and Tamadjan [36]. Theorem 2.4, on the other hand, can be obtained as a corollary to a theorem of Beurling [3, p.155] but the proof given here contains ideas essential for a complete solution to the general approximation problem (cf. Section 3).

THEOREM 2.3. *Let Ω be a bounded simply connected domain having a single rectifiable cut γ and let $\delta(z)$ be the Euclidean distance from z to γ. If $w(z)\le W(\delta(z))$ where W is continuously differentiable and if*

$$(1) \quad t\,\frac{W'(t)}{W(t)} \uparrow +\infty \quad as \quad t \downarrow \infty$$

$$(2) \quad \int_0 \log\log \frac{1}{W(t)}\,dt = +\infty$$

then $H^p(\Omega,wdA)=L_a^p(\Omega,wdA)$ provided w has property (A_p).

THEOREM 2.4. *Let Ω_o be the region obtained by deleting the positive real axis* *from the open unit disk. Let* $\delta(z)$ *be the distance from* z *to the cut. If* $w(z) = W(\delta(z))$ *where*

$$(3) \quad W(t) \downarrow 0 \quad as \quad t \uparrow 0$$

$$(4) \quad \int_0 \log \log \frac{1}{W(t)}\, dt < +\infty$$

then $H^p(\Omega_0, wdA) \neq L^p_a(\Omega_0, wdA)$ *for any* p .

Before commenting on the proofs of these two theorems it is interesting to note that the ideas involved allow us to consider the completeness question for domains with boundary cuts, even when $w \equiv 1$. It is true that $H^p(\Omega, dA) \neq L^p_a(\Omega, dA)$ if Ω has a single "smooth" cut but if there are sufficiently many of them then completeness may indeed occur. The following is a corollary to Theorem 2.3 and generalizes a result of Mergeljan and Tamadjan [28] (cf. also [26, p.120]): Let E be a perfect nowhere dense set of points on the circle $|z| = 1$. Fix a constant ρ, $0 < \rho < 1$, and for each $x \in E$ let $S_x = \{z : \arg z = \arg x, 1-\rho \leq |z| \leq 1\}$ and put $S_E = \cup_{x \in E} S_x$. Thus, $\Omega_E = (|z|<1) \setminus S_E$ is a bounded simply connected domain and $\partial\Omega_E$ consists almost entirely of cuts. Let $CE = (|x| = 1) \setminus E$, $\Delta_t(x) = \{e^{i\theta} : |\theta - \arg x| \leq t\}$ and denote 1-dimensional Lebesgue measure by Λ_1.

CORROLARY 2.5. *The polynomials are complete in* $L^p_a(\Omega_E, dA)$ *for every* p *if there exists a countable set* E' *, everywhere dense in* E, *such that for each* $x \in E'$

(5) $\Lambda_1(\Delta_t(x) \cap CE) \leq W(t)$ *for* $t \leq t(x)$ *and some majorant* W *satisfying condition (1) of Theorem 2.3;*

$$(2) \quad \int_0 \log \log \frac{1}{W(t)}\, dt = +\infty$$

Examples of domains for which these conditions hold can be found in [28].

Proof of Theorem 2.3. Let $\phi : \Omega \to D$ be a conformal map of Ω onto the open unit disk D. Since w has property (A_p), it is enough to verify that $\phi^n(\phi')^{2/p} \in H^p(\Omega, wdA)$ for $n = 0, 1, 2, \ldots$.

To accomplish this we suppose that g is any function in $L^q(\Omega, wdA)$,

$q = p/(p-1)$, with the property that $\int Q\, gwdA = 0$ for every polynomial Q and we prove that $\int \phi^n(\phi')^{2/p}\, gwdA = 0$ for $n = 0, 1, 2, \ldots$. The basic idea goes back to Bers [2] and Havin [15] and consists in verifying the following two assertions:

(a) $\widehat{gw}(z) = \int \frac{gw(\xi)}{\xi - z}\, dA_\xi = 0$ almost everywhere outside of Ω in an appropriate sense;

(b) there exists a sequence of functions $\rho_j \in C_0^\infty(\Omega)$

$$\left\| \frac{\partial \rho_j}{\partial \bar{z}} - gw \right\|_{L^q(\Omega, dA)} \to 0 \quad \text{as} \quad j \to \infty .$$

Once this has been done it follows from (b) that if $F \in L_a^p(\Omega, wdA) \cap L^p(\Omega, dA)$ then

$$\lim_{j \to \infty} \int_\Omega F \frac{\partial \rho_j}{\partial \bar{z}}\, dA = \int_\Omega F\, gwdA .$$

On the other hand, integrating by parts

$$\int_\Omega F \frac{\partial \rho_j}{\partial \bar{z}}\, dA = - \int_\Omega \frac{\partial F}{\partial \bar{z}} \cdot \rho_j\, dA = 0$$

for all j because F is analytic and the support of each ρ_j is away from the boundary. Hence, $\int F\, gwdA = 0$ and, since we can take $F = \phi^n(\phi')^{2/p}$, $n = 0, 1, 2, \ldots$, it follows that $H^p(\Omega, wdA) = L_a^p(\Omega, wdA)$.

Thus, it remains to verify assertions (a) and (b). The first is the crucial point and once it has been established the second follows as a corollary to a theorem of Bagby [1] (cf. also [21, p.313]), at least for $p > 1$. When $p = 1$, however, Bagby's result must be replaced by an argument of Ahlfors (cf. [2, p.3] and [20, p.168]). In any case, we shall comment only on (a) and only to the extent required to give a brief indication of the manner in which the main hypotheses are used. In particular, we shall be content simple to sketch a proof of the fact that $\widehat{gw} \equiv 0$ on the cut γ .

The initial step is to observe that, by virtue of the convexity restriction (1), $\int \delta(z)^{-n} w\, dA < \infty$ for all $n > 0$. Hence $\widehat{gw} \in C^\infty(\gamma)$ and for each $x \in \gamma$

$$\text{(i)} \quad \widehat{gw}^{(n)}(x) = \int_\Omega \frac{gw(z)}{(z-x)^{n+1}}\, dA, \quad n = 0,1,2,\ldots,$$

the differentiation being carried out along γ. It follows that

$$\text{(ii)} \quad \sup_\gamma |\widehat{gw}^{(n)}| \le kn!M_n ,$$

where $M_n = \int \delta(z)^{-(n+1)p} w dA$ and k is some fixed numerical constant. It can also be shown, by repeated applications of an idea of Carleson [11] (cf. also [6, p.126]), that

$$\text{(iii)} \quad \widehat{gw}^{(n)}(x_o) = 0, \quad n = 0,1,2,\ldots .$$

at the point x_o where γ meets $\partial\Omega_\infty$. On the other hand,

$$\text{(iv)} \quad \sum_{n=1}^{\infty} \frac{1}{n(M_n)^{1/n}} = \infty$$

because $\int_o \log\log(1/W(t))dt = +\infty$ and so, by the Denjoy-Carleman Theorem on quasianalyticity, $gw \equiv 0$ on γ. □

Proof of Theorem 2.4. Since W is monotone and since $\int_o \log\log 1/W(t) < +\infty$, there exists a function $W_1 \le W$ with the following properties:

(a) W_1 is continuously differentiable and $t\frac{W_1'(t)}{W_1(t)} \uparrow +\infty$ as $t \downarrow 0$;

(b) $\int_0 \log\log \frac{1}{W_1(t)}\, dt < +\infty$.

Since $(W_1)^2$ also has these properties, we can find for any $\xi \in [0,1]$ another continuously differentiable function $\mu(z)$ everywhere defined and satisfying:

(i) $\mu(\xi) = 1$;

(ii) $\mu \equiv 0$ off a compact subset of the open unit disk;

(iii) $\left|\frac{\partial\mu}{\partial\bar{z}}\right| \le C\, W_1(\delta(z))^2$.

For a proof see Dyn'kin [12, p.188].

If Q is any polynomial then, according to Green's Theorem,

$$Q(\xi) = Q(\xi)\mu(\xi) = \frac{-1}{\pi} \int_{\Omega_o} Q(z)\, \frac{\partial\mu/\partial\bar{z}}{(z-\xi)}\, dA_z$$

and consequently

$$|Q(\xi)| \leq C \int_{\Omega_o} |Q(z)| \frac{w_1(z)^2}{\delta(z)} dA ,$$

where $w_1(z) = W_1(\delta(z))$. Because the convexity condition (a) implies that $w_1(z)/\delta(z)$ is bounded on Ω_o, it is easy to see that

$$|Q(\xi)| \leq C \|Q\|_{L^1(\Omega_o, w_1 dA)}$$

and it follows that the map $Q \to Q(\xi)$ is a bounded linear functional in each of the $L^p(\Omega_o, wdA)$ norms, $p \geq 1$. By a theorem of Sinanjan [34], $H^p(\Omega_o, wdA) \neq L^p_a(\Omega_o, wdA)$ for any p. □

3. THE GENERAL APPROXIMATION PROBLEM

In order to study the completeness problem for the most general regions and weights we shall make extensive use of the notion of a bounded point evaluation. A solution based on this concept seems to have been first advocated by Mergeljan [27] in 1955. He conjectured that for the polynomials to be complete in $L^2_a(\Omega, dA)$ it is necessary and sufficient that for each point $\xi \in \partial\Omega$ and each $\varepsilon > 0$ there exist a polynomial Q with the following properties:

$$(1) \quad \|Q\|_{L^2(\Omega, dA)} < \varepsilon ;$$

$$(2) \quad \sup_{|z-\xi| \leq \varepsilon} |Q(z)| > 1 .$$

This suggests that perhaps $H^2(\Omega, dA) = L^2_a(\Omega, dA)$ if and only if $H^2(\Omega, dA)$ has no bounded point evaluations on $\partial\Omega$. Sinanjan [34], [25, p.726] subsequently conjectured that the latter is indeed the case and he succeeded in establishing necessity.

The following theorem can be found in [8] and it confirms the sufficiency of Sinanjan's conjecture in a more general setting than that in which it was originally posed.

THEOREM 3.1. *Let* Ω *be a bounded simply connected domain and let* $w \in L^\infty$ *be a weight having property* (A_p). *In oder for the polynomials to be complete in* $L^p_a(\Omega, wdA)$ *it is necessary and sufficient that*

$H^p(\Omega, wdA)$ have no bounded point evaluations on $\partial\Omega$.

Although this gives a complete solution to the general approximation problem, it is somewhat unsatisfactory because there is no effective method to determine if $H^p(\Omega, wdA)$ has a bounded point evaluation or not. However, Theorem 3.1 can be used to give a characterization of completeness in terms of a more familiar concept.

THEOREM 3.2. *Let Ω be a bounded simply connected domain and let $w \in L^\infty$ be a weight having property (A_p). If the polynomials fail to be complete in $L_a^p(\Omega, wdA)$ then there exists a point $\xi_o \in \partial\Omega$ and an open set U containing ξ_o such that every function in $H^p(\Omega, wdA)$ admits an analytic continuation to U.*

In case $w \equiv 1$ the continuation phenomenon is implicit in Mergeljan's conjecture [27]. As previously noted, it has also been observed in a wide range of special cases by Keldyš, Šaginjan, Tamadjan, Havin, Beurling and others. For additional information and background material see [25] and [26] and the articles cited therein.

For the sufficiency part of Theorem 3.1 we suppose that $H^p(\Omega, wdA)$ has no bounded point evaluations on $\partial\Omega$. To prove that $H^p(\Omega, wdA) = L_a^p(\Omega, wdA)$ we must show that if $g \in L^q(\Omega, wdA)$ and if $\int Q\, gwdA = 0$ for every polynomial Q then $\int F\, gwdA = 0$ for every F of the form $\phi^n(\phi')^{2/p}$, where ϕ maps Ω conformally onto the open unit disk. As in the proof of Theorem 2.3, it is sufficient to verify that $\widehat{gw} = 0$ almost everywhere on $\partial\Omega$ in an appropriate sense--in this case, almost everywhere with respect to an appropriate capacity.

For each $\lambda > 0$ we let $E_\lambda = \{z : |\widehat{gw}(z)| < \lambda\}$. Under the assumption that ξ_o in not a bounded point evaluation for $H^p(\Omega, wdA)$ it can be shown that almost every circle $|z - \xi_o| = r$ meets E_λ in a set of positive linear measure (cf.[8]). Thus, in this sense, E_λ is thick at every $\xi_o \in \partial\Omega$. Because $\widehat{gw}$ belongs to the Sobolev space $W^{1,q}$, it follows that, except for a set of capacity zero, there exists a sequence $\xi_j \in E_\lambda$ so that $\xi_j \to \xi_o$ and

$$\lim_{j\to\infty} \widehat{gw}(\xi_j) = \widehat{gw}(\xi_o) .$$

Hence, $|\widehat{gw}| \le \lambda$ almost everywhere with respect to capacity on $\partial\Omega$. Since this holds for every $\lambda > 0$, we conclude that $\widehat{gw} = 0$ "a.e" on $\partial\Omega$.

The proof of Theorem 3.2 goes as follows: If $H^p(\Omega,wdA) \ne L^p_a(\Omega,wdA)$ then $H^p(\Omega,wdA)$ has a bounded point evaluation at some $\xi_o \in \partial\Omega$. For this point it can be shown that there is an open set U and fixed constant C such that

$$|Q(\xi)| \le C \, \|Q\|_{L^p(\Omega,wdA)}$$

for every polynomial Q and every $\xi \in U$. Here, of course, $\xi_o \in U$. This ensures that every $f \in H^p(\Omega,wdA)$ has an analytic continuation to U. □

REFERENCES

[1] T. Bagby, Quasi-topologies and rational approximation, *J. Functional Analysis* 10(1972), 259-268.

[2] L. Bers, An approximation theorem, *J. Analyse Math.* 14(1965), 1-14

[3] A.Beurling, Analytic continuation across a linear boundary, *Acta Math.* 128(1972), 153-182.

[4] A. Beurling, Quasinalyticity and general distributions, *Lecture Notes, Stanford Univ.* (1961).

[5] J. Brennan, Invariant subspaces and weighted polynomial approximation, *Ark. Mat.* 11(1973), 167-189.

[6] J. Brennan, Approximation in the mean by polynomials on non-Carathéodory domains, *Ark. Mat.* 15(1977), 117-168.

[7] J. Brennan, The integrability of the derivative in conformal mapping, *J. London Math. Soc.* 18(1978), 261-272.

[8] J. Brennan, Point evaluations, invariant subspaces and approximation in the mean by polynomials, *J. Functional Analysis* (to appear).

[9] J. Brennan, Invariant subspaces and subnormal operators, *Proc. Symp. Pure Math.* (to appear).

[10] T. Carleman, Über die Approximation analytischer Funktionen durch lineare Aggregate von vorgegebenen Potenzen, *Ark. Mat. Astr. Fys.* 17(1923), 1-30.

[11] L. Carleson, Mergeljan's theorem on uniform polynomial approximation, *Math. Scand.* 15(1965), 167-175.

[12] E.M. Dyn'kin, Functions with a given estimate for $\partial f/\partial \bar{z}$ and N. Levinson's theorem, *Math. USSR Sbornik* 18(1972), 181-189; *Mat. Sb.* 89(1972), 182-190.

[13] M.J. Džrbašjan, Metric theorems on completeness and the representability on analytic functions, Thesis, Erevan (1948).

[14] O.J. Farrell, On approximation to an analytic function by polynomials, *Bull. Amer. Math. Soc.* 40(1934), 908-914.

[15] V.P. Havin, Approximation in the mean by analytic functions, *Soviet Math Dokl.* 9(1968),245-248; *Dokl. Akad. Nauk SSSR* 178(1968), 1025-1028.

[16] V.P. Havin, Approximation in the mean by polynomials on certain non-Carathéodory domains, I & II, *Izv. Vyss Ucebn. Zaved. Mat.* 76(1968), 86-93 & 77(1968), 87-94.

[17] V.P. Havin and V.G. Maz'ja, Approximation in the mean by analytic functions, *Vestnik Leningrad Univ.* 13(1968), 64-74.

[18] V.P. Havin and V.G. Maz'ja, Applications of (p, ℓ)-capacity to some problems in the theory of exceptional sets, *Math. USSR Sbornik* 19 (1973), 547-580; *Mat. Sb.* 90(1973), 558-591.

[19] L.I. Hedberg, Weighted mean approximation in Carathéodory regions, *Math. Scand.* 23(1968), 113-122.

[20] L.I. Hedberg, Approximation in the mean by analytic functions, *Trans. Amer. Math. Soc.* 163(1972), 157-171.

[21] L.I. Hedberg, Non-linear potentials and approximation in the mean by analytic functions, *Math. Z.* 129(1972), 299-319.

[22] M.V. Keldyš, Sur l'approximation en moyenne quadratique des fonctions analytiques, *Mat. Sb.* 47(1939), 391-401.

[23] M.V. Keldyš, Sur l'approximation en moyenne par polynômes des fonctions d'une variable complexe, *Mat. Sb.* 58(1945), 1-20.

[24] A.I. Markuševič, Conformal mappings of regions with variable boundaries and applications to the approximation of analytic functions by polynomials, Thesis, Moscow (1934).

[25] M.S. Melnikov and S.O. Sinanjan, Questions in the theory of approximation of functions of one complex variable, in *Contemporary Problems of Mathematics,* vol.4, Itogi Nauki i Tekhniki, VINITI, Moscow, (1975), 143-250; English translation, *J. Soviet Math.* 5(1976), 688-752.

[26] S.N. Mergeljan, On the completeness of systems of analytic functions, *Amer. Math. Soc. Translations* 19(1962), 109-166; *Uspehi Mat. Nauk* 8(1953), 3-63.

[27] S.N. Mergeljan, General metric criteria for the completeness of systems of polynomials, *Dokl. Akad. Nauk, SSSR* 105(1955), 901-904.

[28] S.N. Mergeljan and A.P. Tamadjan, On completeness in a class of non-Jordan regions. *Amer. Math. Soc. Translations* 35(1964), 79-94; *Izv. Akad. Nauk. Armjan. SSR* 7(1954), 1-17.

[29] A.L. Šaginjan, On a criterion for incompleteness of a system of analytic functions, *Dokl. Akad. Nauk. Armjan SSR* 5(1946), 97-100.

[30] A.L. Šaginjan, A problem in the theory of approximation in the complex domain, *Sibirsk Mat. Z.* 1(1960), 523-543.

[31] H.S. Shapiro, Weighted polynomial approximation and boundary behavior of analytic functions, in *Contemporay Problems in the Theory*

of Analytic Functions, Nauka, Moscow (1966), 326-335.

[32] H.S. Shapiro, Some remarks on weighted polynomial approximations of holomorphic functions, *Math. USSR Sbornik* 2(1967), 285-293; *Mat. Sb.* 73(1967), 320-330.

[33] S.O. Sinanjan, Approximation by polynomials and analytic functions in the areal mean, *Amer. Math. Soc. Translations* 74(1968), 91-124; *Mat. Sb.* 69(1966), 546-578.

[34] S.O. Sinanjan, Approximation by polynomials in the mean with respect to area, *Math. USSR Sbornik* 11(1970), 411-421; *Mat. Sb.* 82(1970), 444-455.

[35] S.O. Sinanjan, On the completeness of the polynomials in the space L^p, *Mat. Zametki* 24(1978), 73-83.

[36] A.P. Tamadjan, A theorem of M.V. Keldys, *Izv. Akad. Nauk. Armjan. SSR* 6(1953), 5-11.

Department of Mathematics
University of Kentucky
Lexington, Kentucky 40506.

QUELQUES PROPRIÉTÉS DU PRÉDUAL DE H^∞

par

Jacques Chaumat

Soit D le disque unité du plan complexe, T le tore unité du plan complexe, m la mesure de Lebesgue sur T. Soit A(D) l'algèbre des fonctions holomorphes dans D et continues sur $D \cup T$, H^∞ l'algèbre des fonctions holomorphes et bornées dans D. On peut considérer A(D) comme une algèbre uniforme sur T et identifier H^∞ à l'adhérence de A(D) dans $L^\infty(m)$ pour la topologie $\sigma(L^\infty(m), L^1(m))$.

THÉORÈME(J.-P. Kahane [20]). *Soit* $\{f_n\}_{n\in\mathbb{N}}$ *une suite de fonctions de* $L^1(m)$ *telles que pour toute fonction* g *de* H^∞ *la suite* $\{\int f_n g\, dm\}_{n\in\mathbb{N}}$ *converge. Alors il existe une fonction* f *de* $L^1(m)$ *telle que* $\lim_{n\to+\infty} \int f_n\, g\, d_m = \int f\, g\, dm$ *pour toute fonction* g *de* $A(D)$.

L'argument de la preuve repose sur la caractérisation bien connue des ensembles pics de A(D): un sous-ensemble fermé E de T est un ensemble pic A(D) si et **seulement** si $m(E) = 0$ [18].

DÉFINITION. On dit que E un sous-ensemble fermé de T est pic pour A(D) si et seulement si il existe une fonction g de A(D) vérifiant $g(x)=1$ pour tout x de E et $|g(x)|<1$ pour tout x de $T\setminus E$.

Le but de cet exposé est de donner différents prolongements du théorème de J.-P. Kahane.

I. QUELQUES PROPRIÉTÉS DE $L^1(m)/H^{\infty\perp}$

THÉORÈME(M. Mooney [23]). *Supposons que l'on ait* $\{f_n\}_{n\in\mathbb{N}}$ *une suite de fonctions de* $L^1(m)$ *telle que pour toute fonction* g *de* H^∞

la suite $\{\int f_n\, g\, dm\}_{n\in\mathbb{N}}$ *converge. Alors il existe une fonction f de* $L^1(m)$ *telle que* $\lim_{n\to+\infty} \int f_n\, g\, dm = \int f\, g\, dm$ *pour toute fonction g de* H^∞.

Considérons l'orthogonal $H^{\infty\perp}$ de H^∞ dans $L^1(m)$. Alors $L^1(m)/H^{\infty\perp}$ a pour dual H^∞ et on peut interpréter le résultat de M. Mooney en écrivant que $L^1(m)/H^{\infty\perp}$ est faiblement séquentiellement complet.

Notons M le spectre de $L^\infty(m)$, muni de la topologie de Gelfand. La transformation de Gelfand est un isomorphisme isométrique de $L^\infty(m)$ sur l'algèbre $C(M)$ des fonctions continues sur M à valeurs complexes: $\hat{g}$ désigne la transformée de Gelfand d'une fonction g de $L^\infty(m)$. L'algèbre H^∞ s'identifie à une sous-algèbre fermée $\hat{H}^\infty$ de $C(M)$. Une forme linéaire ℓ sur $L^\infty(m)$ continue, peut être représentée de manière unique par une mesure $\hat{\ell}$ borélienne bornée sur M; en particulier la mesure de Lebesgue m peut être représentée par une mesure $\hat{m}$ sur M. ([13] pp. 17-19).

THÉORÈME (E. Amar et A. Lederer [2]). *Soit G un fermé de M de* $\hat{m}$ *mesure nulle. Alors il existe un fermé G' de M, de* $\hat{m}$ *mesure nulle, contenant G et pic pour* $\hat{H}^\infty$.

En utilisant ce résultat et en reprenant l'argument de J.-P. Kahane, E. Amar [1] donne une preuve simple du théorème de M. Mooney.

Dans ce paragraphe, on présente d'autres propriétés du couple d'espaces vectoriels $(L^1(m)/H^{\infty\perp}, H^\infty)$ découlant du théorème de E. Amar et A. Lederer. Certaines d'entre elles présentent une certaine analogie avec des propriétés connues du couple $(L^1(m), L^\infty(m))$.

Pour une fonction g de H^∞ considérons l'élément Tg de $L^1(m)/H^{\infty\perp}$ défini par $\langle Tg, g'\rangle = \int \bar{g}\, g'\, dm$ pour toute fonction g' de H^∞. Remarquons que, (avec des notations de normes évidentes) si $\|g\|_\infty = 1$ alors $\|g\|_1^2 \leq \|Tg\|_{L^1(m)/H^{\infty\perp}} \leq \|g\|_1$. En conséquence T est une application linéaire continue, injective et d'image dense de H dans $L^1(m)/H^{\infty\perp}$.

DÉFINITION. Une suite bornée $\{f_n\}_{n\in\mathbb{N}}$ dans $L^1(m)/H^{\infty\perp}$ est d'interpolation si et seulement si, pour toute suite bornée, $\{\alpha_n\}_{n\in\mathbb{N}}$, de nombres complexes, il existe une fonction g de H^∞ telle que $\langle f_n, g\rangle = \alpha_n$ pour tout n.

On a alors le théorème suivant analogue au théorème classique de Dunford-Pettis caractérisant les parties faiblement relativement compactes de $L^1(m)$.

THÉORÈME 1 [6]. *Pour une partie K de $L^1(m)/H^{\infty\perp}$, les propositions suivantes sont équivalentes:*

1) K est faiblement relativement compacte;

2) Pour tout $\varepsilon>0$, il existe $\delta>0$ tel que $g\in H^\infty$, $\|g\|_\infty\le 1$ et $\|Tg\|_{L^1(m)/H^{\infty\perp}} \le\delta$ implique $\sup_{f\in K}|\langle f,g\rangle|\le\varepsilon$;

3) $\lim_{C\to+\infty} [\sup_{f\in K} (\inf_{\substack{g\in H^\infty \\ \|g\|_\infty\le C}} \|f-Tg\|_{L^1(m)/H^\infty}] = 0$;

4) K est bornée et ne contient aucune suite d'interpolation.

Les idées développées dans la preuve de ce théorème [6] se rattachent à l'argument de J.-P. Kahane et E. Amar. Les fonctions "pic" de H^∞ peuvent jouer le rôle des fonctions caractéristiques de L^∞. Signalons que V.P. Havin [16] obtient indépendamment des résultats proches du théorème et une autre preuve du théorème de M. Mooney.

THÉORÈME (F. Delbaen et S.V. Kisliakov[11] et [22]). *Une partie K faiblement relativement compacte de $L^1(m)/H^{\infty\perp}$ est le quotient par $H^{\infty\perp}$ d'une partie K' faiblement relativement compacte de $L^1(m)$.*

Notons pour une partie K bornée de $L^1(m)/H^{\infty\perp}$ et $\varepsilon>0$ $\eta(H^\infty,K,\varepsilon) = \sup\{|\langle f,g\rangle|,\ f\in K,\ g\in H^\infty\ \|g\|_\infty\le 1$ et $\|Tg\|_{L^1(m)/H^{\infty\perp}}\le\varepsilon\}$, $\eta(H^\infty,K) = \lim_{\varepsilon\to 0}\eta(H^\infty,K,\varepsilon)$. Cette limite existe car la fonction $\varepsilon\to\eta(H^\infty,K,\varepsilon)$ est positive et décroissante.

Cette notion, introduite pour $(L^1(m),L^\infty(m))$ par M.Kadec et A. Pełczyński [21], mesure le "défaut" de compacité de K. On peut récrire le théorème de F. Delbaen et S.V. Kisliakov de la manière suivante:

Une partie K bornée $L^1(m)/H^{\infty\perp}$ vérifiant $\eta(H^\infty,K)=0$ est le quotient par $H^{\infty\perp}$ d'une partie K' de $L^1(m)$ vérifiant $\eta(L^\infty,K)=0$.

On a alors le théorème suivant dont la preuve approfondit l'argument de F. Delbaen et S.V. Kisliakov.

THÉORÈME 2 [6]. *Une partie K bornée de $L^1(m)/H^{\infty\perp}$ est le quotient par $H^{\infty\perp}$ d'une partie K' bornée de $L^1(m)$ vérifiant $\eta(L^\infty,K') = \eta(H^\infty,K)$.*

Donnons quelques indications sur la preuve. Pour une fonction f de $L^1(m)$ il existe une fonction f' de $H^{\infty\perp}$ telle que $\|f-f'\|_{L^1(m)} = \inf_{h\in H^{\infty\perp}} \|f-h\|_{L^1(m)}$. En conséquence, pour tout élément ℓ de $L^1(m)/H^{\infty\perp}$ il existe une fonction ℓ' de $L^1(m)$ telle que $\|\ell\|_{L^1(m)/H^{\infty\perp}} = \|\ell'\|_{L^1(m)}$ et $\ell'/H^{\infty\perp} = \ell$. On peut alors vérifier que pour une partie K bornée de $L^1(m)/H^{\infty\perp}$ la partie K' de $L^1(m)$ définie par $K' = \{\ell'\in L^1(m),\ \ell\in K\}$ convient.

DÉFINITIONS. Un espace de Banach E est appelé un prédual isométrique d'un espace de Banach F si le dual E* de E est isométriquement isomorphe à F.

Un espace de Banach E est appelé prédual isométrique unique d'un espace de Banach F si tout prédual isométrique de F est isométriquement isomorphe à E. On dira aussi que F est un espace de Banach à prédual isométrique unique.

L'espace L^1 d'une mesure finie est le prédual isométrique unique de L^∞ [15]. Récemment T. Ito a montré que tout sous espace fermé de L^1 d'une mesure finie est le prédual isométrique unique de son dual [19].

Résolvant une question de P. Porcelli, P. Wojtasczyck [24] et T. Ando [3] ont montré indépendamment que H^∞ est un espace de Banach à prédual isométrique unique.

THEOREME 3 [7]. *Tout sous espace fermé de $L^1(m)/H^{\infty\perp}$ est le prédual isométrique unique de son dual.*

La preuve s'apparente à celle de P. Wojtasczyck [24].

Une théorie analogue peut être développée dans un cadre abstrait [6] : soit (S,Σ,m) un espace de probabilité et H^∞ une sous-algèbre de $L^\infty(m)$ fermée pour la topologie $\sigma(L^\infty(m),L^1(m))$ et contenant les constantes. On suppose que H^∞ a "assez d'ensembles pics"[i.e. la conclusion du théorème de E. Amar et A. Lederer est satisfaite]. Alors on a les théorèmes 1, 2 et 3.

Ainsi $L^\infty(m)$ a "assez d'ensembles pics" et on peut redonner des preuves de théorèmes classiques sans faire jouer aux fonctions caractéristiques un rôle particulier.

I. Cnop et F. Delbaen [9] décrivent une large classe d'algèbres ayant "assez d'ensembles pics".

On donne maintenant des exemples d'algèbres n'ayant pas tout à fait assez d'ensembles pics" et pour lesquelles les théorèmes 1, 2 et 3 sont vrais [6].

Soit U un ouvert du plan complexe $\mathbb{C}$ obtenu en enlevant au disque $D=\{z\in\mathbb{C}:|z|<1\}$ une suite de disques $\bar{D}_n=\{z\in\mathbb{C}:|z-x_n|\le r_n\}$, pour $n=1,2,...$ centrés sur le demi-axe réel positif, s'accumulant en 0, deux à deux disjoints et vérifiant $\sum_{n=1}^{+\infty}\frac{r_n}{x_n}<+\infty$. On dit que U est un ouvert de type L. [25] .

Notons m la mesure de Lebesgue sur le bord de U et H^∞ l'algèbre des fonctions holomorphes et bornées dans U. On peut considérer H^∞ comme une sous-algèbre de $L^\infty(m)$ fermée pour la topologie $\sigma(L^\infty(m), L^1(m))$.

Un homomorphisme de l'algèbre H^∞ joue un rôle particulier: c'est l'unique homorphisme ϕ_0 de H^∞ continu pour la topologie $\sigma(L^\infty(m), L^1(m))$ et vérifiant $\phi_0(z)=0$. On l'appelle homomorphisme distingué [25], [14].

C'est l'étude de l'homomorphisme distingué qui permet d'une part de montrer que H^∞ ne vérifie pas la conclusion du théorème de E. Amar et A. Lederer, et d'autre part de prouver les théorèmes 1, 2 et 3 [6] .

II. QUELQUES EXTENSIONS DU THÉORÈME DE J.-P. KAHANE

On peut présenter le théorème de J.-P. Kahane dans le cadre suivant. Soit A une algèbre uniforme sur un compact X et μ une mesure de probabilité sur X; soit $H^\infty(\mu)$ l'adhérence de A dans $L^\infty(\mu)$ pour la topologie $\sigma(L^\infty(\mu),L^1(\mu))$. On dira que le couple (A,μ) vérifie la propriété K si la conclusion du théorème de J.-P. Kahane est satisfaite.

THÉORÈME (E. Heard [17]). *Supposons que toute mesure orthogonale à l'algèbre A soit absolument continue par rapport à la mesure μ. Alors le couple (A,μ) vérifie la propriété K.*

E. Heard démontre ce théorème dans le cadre des espaces vectoriels uniformes. Dans le cas de l'algèbre A(D) et de la mesure de Lebesgue m, l'hypothèse est satisfaite d'après le théorème de F. et M. Riesz [18]. La preuve de théorème consiste à montrer qu'un fermé de X de μ-mesure nulle est un ensemble pic pour l'algèbre A. On conclut alors en utilisant l'argument de J.-P. Kahane.

On peut espérer obtenir la propriété K pour un couple (A,μ) chaque fois que l'on aura un analogue du théorème de F. et M. Riesz. Le but de ce paragraphe est de décrire quelques exemples.

Le cas d'algèbres uniformes sur des compacts du plan complexe a été étudié dans [5] . On obtient les théorèmes suivants.

Soit X un compact du plan complexe et R(X) l'algèbre des limites uniformes sur X de fractions rationnelles d'une variable complexe à pôles hors de X. Soit μ une mesure de probabilité sur X.

THEOREME 4 [5]. *Le couple $(R(K),\mu)$ vérifie la propriété K.*

Soit U un ouvert borné du plan complexe et A(U) l'algèbre de fonctions holomorphes dans U est continues sur l'adhérence $\bar{U}$ de U. Soit μ une mesure de probabilité sur $\bar{U}$.

THEOREME 5 [5]. *Le couple $(A(U),\mu)$ vérifie la propriété K.*

Etudions maintenant quelques exemples d'algèbres uniformes de plusieurs variables complexes.

Soit U un domaine de $\mathbb{C}^n$ borné strictement pseudoconvexe à frontière ∂U de classe C^2 [[10] la boule unité de $\mathbb{C}^n$ est un exemple d'un tel domaine]. On note σ la mesure superficielle sur ∂U et A(U) l'algèbre des fonctions holomorphes dans U et continues sur $U \cup \partial U$. On peut considérer A(U) comme une algèbre uniforme sur ∂U. On suppose que 0 appartient à U.

DÉFINITION. Une mesure positive η sur ∂U est représentative de 0 pour A(U) si et seulement si pour toute fonction f de A(U) on a $\int f\, d\eta = f(0)$.

THÉORÈME (B. Cole. R.M. Range et G. Henkin [10]). *Une mesure sur ∂U orthogonale à $A(U)$ est absolument continue par rapport à une mesure représentative de 0.*

Dans le cas de $A(D)$ on sait que la mesure de Lebesgue m est l'unique mesure représentative de 0; on retrouve alors le théorème de F et M. Riesz [18]. En utilisant des décompositions de mesures ([13] pp. 40-46) et des constructions d'ensembles pics([13] pp. 56-60) on peut montrer le théorème suivant.

THÉORÈME 6 [8]. *Le couple $(A(U),\sigma)$ vérifie la propriété K.*

Soit D^2 le polydisque unité de $\mathbb{C}^2$ [i.e. $D^2 = \{(z,w)\in\mathbb{C}^2, |z|<1 |w|<1\}$. Soit m^2 la mesure de Lebesgue sur $T^2=\{(z,w)\in\mathbb{C}^2, |z| = |w| = 1\}$. Soit $A(D^2)$ l'algèbre des fonctions holomorphes dans D^2 et continues sur $\bar{D}^2 = \{(z,w)\in\mathbb{C}^2, |z| \leq 1, |w|\leq 1\}$. On peut considérer $A(D^2)$ comme une algèbre uniforme sur T^2.

Pour une mesure α sur T^2 on note α_1 la mesure définie sur T par $\alpha_1(E) = |\alpha|(E\times T)$ pour tout borélien E de T. On définit α_2 de manière analogue: $\alpha_2(E) = |\alpha|(T\times E)$ pour tout borélien E de T.

THÉORÈME (Bui Doan Khanh [4]). *Soit γ une mesure sur T^2 orthogonale à $A(D^2)$ telle que les mesures γ_1 et γ_2 soient absolument continues par rapport à la mesure de Lebesgue m sur T. Alors γ est absolument continue par rapport à une mesure représentative de 0.*

Remarquons qu'une mesure sur T^2 orthogonale à $A(D^2)$ n'est pas nécessairement absolument continue par rapport à une mesure représentative de 0. Néanmoins on a le théorème suivant.

THÉORÈME 7 [8]. *Le couple $(A(D^2),m^2)$ vérifier la propriété K.*

Considérons le polydisque infini $\bar{D}^\infty = \prod_{i=1}^{\infty} \bar{D}_i$ avec $\bar{D}_i = \bar{D}$, pour tout i, muni de la topologie produit. C'est un espace topologique compact. Soit $A(\bar{D}^\infty)$, l'algèbre uniforme sur $\bar{D}^\infty$ engendrée par les polynômes d'un nombre fini de variables complexes. Soit m^∞ la masse de Lebesgue produit sur $\bar{D}^\infty$.

THÉORÈME 8 [8] . *Le couple* $(A(\bar{D}^\infty), m^\infty)$ *ne vérifie pas la propriété K.*

Considérons pour chaque i, la forme linéaire Π_i définie sur les polynômes et à valeurs complexes par $\Pi_i(P)$ = le coefficient du monôme z_i. Cette forme linéaire Π_i se prolonge de manière continue à $A(\bar{D}^\infty)$ et, de manière continue pour la topologie $\sigma(L^\infty(m^\infty), L^1(m^\infty))$, à $H^\infty(m^\infty)$. De plus on a, pour toute fonction f de $H^\infty(m^\infty)$, $\sum_{i=1}^{\infty} |\Pi_i(f)| \leq \|f\|_\infty$.

[On utilise ici une remarque communiquée à l'auteur par J.-P. Rosay].

De là on déduit que la suite $(\sum_{i=1}^{n} \Pi_i)_{n\in\mathbb{N}}$ de formes linéaires sur $H^\infty(m^\infty)$ converge ponctuellement, et on peut vérifier que la limite $\sum_{i=1}^{+\infty} \Pi_i$ ne peut pas être représentée sur $A(\bar{D}^\infty)$ par une fonction de $L^1(m^\infty)$.

Signalons pour conclure ce paragraphe que T.W. Gamelin a étendu dans un cadre plus abstrait le théorème de J.-P. Kahane [12].

III. QUELQUES QUESTIONS

Soit U un ouvert du plan complexe et $H^\infty(U)$ l'algèbre des fonctions holomorphes et bornées dans U. Le prédual isométrique de $H^\infty(U)$ est-il unique? Est-il faiblement séquentiellement complet?

On peut se poser les mêmes questions pour un ouvert strictement pseudoconvexe de $\mathbb{C}^n$ ou pour le polydisque de $\mathbb{C}^n$.

BIBLIOGRAPHIE

[1] E. Amar, Sur un théorème de Mooney relatif aux fonctions analytiques bornées. *Pacific J. Math.* 16 (1973), 191-194.

[2] E. Amar, E et A. Lederer.. Points exposées de la boule unité de $H^\infty(D)$. *C.R. Acad. Sc. Paris* 272 (1971), 1449-1452.

[3] T. Ando. Uniqueness of predual of H^∞. Preprint.

[4] Bui Doan Khanh. Les A-mesures sur le tore de dimension 2 et le calcul symbolique de deux contractions permutables. *J. Funct. Anal.* 15 (1974), 33-44.

[5] J. Chaumat. Adhérence faible étoile d'algèbres de fractions ra-

tionnelles. *Thèse Univ. Paris-Sud*, 1975.

[6] J. Chaumat. Quelques propriétés du couple d'espaces vectoriels $(L^1(m)/H^{\infty\perp}, H^\infty)$. *Prépublication du département des mathématiques, Univ. Paris-Sud.*

[7] J. Chaumat. Unicité de prédual. *Preprint.*

[8] J. Chaumat. Non publié.

[9] I. Cnop, et F. Delbaen. A Dunford-Pettis theorem for $L^1/H^{\infty\perp}$. *Wrije Universiteit, Brussel (1975). Preprint.*

[10] B. Cole, and R.M. Range. A measures on complex manifolds and some applications. *J. Funct. Anal.* 11 (1972), 393-400.

[11] F. Delbean. Weakly compact sets in L^1/H^1_0. *Universiteit, Brussel* (1975). *Preprint.*

[12] T.W. Gamelin. An abstract Kahane theorem (1975). *Preprint.*

[13] T.W. Gamelin. *Uniform algebras.* Prentice Hall series in Modern Analysis (1969).

[14] T.W. Gamelin, and J. Garnett. Distinguished homomorphism and Fiber algebra. *Amer. J. Math.* 42 (1970), 455-474.

[15] A. Grothendieck. Caractérisation vectorielle métrique des espaces L^1. *Canadian J. Math.* (1955), 552-561.

[16] V.P. Havin. The spaces H^∞ and L^1/H^1_0 (Investigations on linear operators and theory of functions IV). *Zap. Naucn. Sem. Leningrad Otdel. Math. Inst. Steklov* 39 (1974), 120-140. (en russe).

[17] E.Heard. A sequential F. and M. Riesz theorem. *Proc. Amer. Math. Soc.* 18 (1967), 832-835.

[18] K. Hoffman. *Banach spaces of analytic functions.* Prentice Hall series in modern analysis (1962).

[19] T. Ito. On uniqueness of preduals. *Notices Amer. Math. Soc.* 25

(1978), A-294.

[20]J.-P. Kahane. Another theorem on bounded analytic functions. *Proc. Amer. Math. Soc.* 18 (1967), 827-831.

[21]M.I. Kadec, and A. Pełczyński. Bases, Lacunary sequences and complemented subspaces in the spaces L_p. *Studia Math.* 2 (1962), 161-179.

[22]S. V. Kisliakov. On the conditions of Dunford-Pettis, Pełczyński and Grothendieck. *Dokl. Akad. Nauk. SSSR* 225 (1975), 1252-1255.

[23]M. Mooney. A theorem on bounded analytic functions. *Pacific J. Math.* 43 (1972), 457-462.

[24]P. Wojtasczyck. On projections in spaces of analytic functions with applications. *Warszawa* 1977. *Preprint.*

[25]L. Zalcman. Bounded analytic functions on domains of infinite connectivity. *Trans. Amer. Math. Soc.* 144 (1969), 241-270.

Université de Paris-Sud
Orsay

ENSEMBLES DE ZÉROS, ENSEMBLES PICS POUR A(D) ET $A^\infty(D)$

par

Anne-Marie Chollet

1. D désigne un domaine borné de $\mathbb{C}^n$, strictement pseudoconvexe à frontière ∂D de classe C^2. Il existe donc une fonction r, à valeurs réelles, de classe C^2 dans un voisinage de $\bar{D}$ l'adhérence de D telle que

i) $D=\{z\in\mathbb{C}^n;\ r(z)<0\}$

ii) grad $r\neq 0$ sur ∂D,

iii) r est strictement plurisousharmonique dans un voisinage de ∂D.

2. On note

A(D) la classe des fonctions analytiques dans D, continues dans $\bar{D}$,
$A^\infty(D)$ la classe des fonctions analytiques dans D, continues ainsi que leurs dérivées de tous ordres dans $\bar{D}$.

Dans tout ce qui suit, E est un sous-ensemble fermé de ∂D. On se propose pour chacune des classe décrites ci-dessus d'étudier les ensembles de zéros et les ensembles pics.

E est un ensemble de zéros pour une classe A s'il existe une fonction f de A telle que

$$E=\{z\in\bar{D}\ ;\ f(z)=0\}.$$

E est un ensemble pic pour A s'il existe une fonction f de A telle que

$$f=1 \quad \text{sur} \quad E \quad \text{et} \quad |f|<1 \quad \text{dans} \quad \bar{D}\setminus E$$

ou, ce qui est équivalent, s'il existe une fonction g de A telle que

$$g=0 \quad \text{sur} \quad E \quad \text{et} \quad \operatorname{Re} g>0 \quad \text{dans} \quad \bar{D}\setminus E\ .$$

Lorsque A est la classe A(D) ces notions sont équivalentes. Il n'en est plus de même pour $A^{\infty}(D)$.

ÉTUDE DE A(D).

3. DÉFINITIONS. E est un ensemble d'interpolation pour A(D) si, pour toute fonction f continue sur E, il existe une fonction F de A(D) telle que $F|_E = f$.

E est un ensemble pic-interpolation pour A(D) s'il est à la fois ensemble pic et d'interpolation pour A(D).

4. THÉORÈME [2]. *Pour un sous-ensemble fermé de* ∂D, *la frontière d'un domaine borné strictement pseudoconvexe de* $\mathbb{C}^n$, *les notions: ensemble de zéros, ensemble pic, ensemble d'interpolation, ensemble pic-interpolation pour* A(D) *sont équivalentes.*

Si D est le disque unité du plan complexe, le théorème de Rudin-Carleson [11] donne une caractérisation de tels ensembles: ce sont les fermés de mesure de Lebesgue nulle sur le cercle. Dans le cas de la boule unité de $\mathbb{C}^n$ [19], L. Stout a établi l'équivalence entre les notions ensemble de zéros et ensemble pic-interpolation. Le théorème est prouvé par R.E. Valskii [22] pour des domaines étoilés à frontière de classe C^3 et par G. M Henkin et E. M. Čirka [9] pour des domaines simplement connexes.

Dans le cas d'un domaine D strictement pseudoconvexe, certaines de ces équivalences se déduisent d'une application à A(D) de théorèmes généraux sur les algèbres uniformes qui impliquent, entre autres, que la réunion de deux ensembles pics pour A(D) est un ensemble pic pour A(D).

La partie originale du théorème 4 réside dans la preuve de l'implication

$$E \text{ zéro} \Longrightarrow E \text{ pic}.$$

Dans le cas où D est un domaine simplement connexe, si f est une fonction nulle seulement sur E dont le module est strictement inférieur à 1 sur $\bar{D}$, on peut exhiber une fonction de A(D), par exemple $g = \frac{1}{2}(1 - \frac{\log f}{1-\log f})$ qui vaut 1 sur E et dont le module est strictement inférieur à 1 dans $\bar{D} \setminus E$.

Dans le cas général, en chaque point ζ de ∂D, on construit un domaine Δ_ζ strictement pseudoconvexe, simplement connexe, à frontière C^2 tel que Δ_ζ soit inclus dans D et $\partial D \cap \bar{\Delta}_\zeta$ contienne un voisinage de ζ dans ∂D. Il existe alors un voisinage U_ζ de ζ tel que le fermé $\bar{\Delta}_\zeta \cap \partial D$ contienne un voisinage dans ∂D de $F_\zeta = U_\zeta \cap E$ et tel que F_ζ soit un ensemble pic pour $A(\Delta_\zeta)$. On utilise alors les résultats de R. Kerzman [12], B. Cole et M. Range [5] concernant l'opérateur $\bar{\partial}$ pour établir que F_ζ est un ensemble pic pour A(D). Un argument de compacité permet alors de conclure.

5. THÉORÈME [2]. *Soit* D *un domaine borné strictement pseudo-convexe de* $\mathbb{C}^n$ *et* E *un ensemble de zéros pour* A(D), *alors il existe un domaine* $\mathcal{D}$ *strictement pseudoconvexe contenant* $\bar{D} \setminus E$ *et tel que* E *soit un ensemble de zéros pour* $A(\mathcal{D})$.

Si D est le disque unité du plan complexe et E un fermé de mesure nulle sur le cercle, on s'assure que le théorème est vérifié en construisant explicitement une fonction de A(D) nulle seulement sur E et analytique dans un voisinage de $\bar{D} \setminus E$.

Le théorème 5 apporte une réponse à une question posée par W. Rudin.

La preuve du théorème repose sur le lemme suivant et utilise des techniques développées par M. Range [17] et J. Détraz [7].

LEMME. *Soit* D *un domaine borné strictement pseudo-convexe de* $\mathbb{C}^n$, E *un sous-ensemble fermé de* ∂D *et* f *une fonction de* A(D) *nulle seulement sur* E. *Alors, quel que soit* K *sous-ensemble fermé de* $\partial D \setminus E$, *quel que soit* V *voisinage de* K *disjoint de* E *et quel que soit* $\epsilon > 0$, *il existe un domaine* D_1 *strictement pseudo-convexe tel que l'on ait*

$$D_1 \supset D \cup K \quad \text{et} \quad D_1 \setminus D \subset V$$

et tel qu'il existe une fonction ϕ *analytique dans* D_1, *continue dans* $\bar{D}_1 \setminus E$ *vérifiant*

$$||\frac{1}{f}-\phi||_{A(D)}\leq\varepsilon.$$

6. **REMARQUE.** D'après le théorème 4, des résultats analogues s'énoncent pour des sous-ensembles fermés de ∂D qui sont des ensembles pics ou des ensembles fermés de ∂D qui sont des ensembles pics ou des ensembles d'interpolation pour $A(D)$. On peut néanmoins obtenir ces résultats sans recourir au théorème 4 en adaptant dans chaque cas la preuve du théorème 5.

EXEMPLES D'ENSEMBLES PICS POUR A(D).

7. DÉFINITIONS. $\partial D = \{z\in\mathbb{C}^n ; r(z) = 0\}$ est une hypersurface de $\mathbb{C}^n$. On note $T_z(\partial D)$ l'espace tangent en z à ∂D de dimension réelle 2n-1 et grad r(z) le gradient de r évalué en z.

On désigne par J l'opérateur $\mathbb{C}$-linéaire sur $\mathbb{R}^{2n}$ qui correspond à la multiplication par i et on pose, pour tout z de ∂D,

$$\chi_z = \text{grad } r(z) \quad \text{et} \quad \tau_z = J\chi_z.$$

On désigne par $T_z^c(\partial D)$ l'espace tangent complexe, c'est-à-dire le sous-espace complexe maximal de l'espace tangent réel $T_z(\partial D)$. Il est de dimension complexe n-1 et on a la décomposition orthogonale réelle, en tout point z de ∂D

$$T_z(\partial D) = \mathbb{R}[\tau_z] \oplus T_z^c(\partial D)$$

où $\mathbb{R}[\tau_z]$ est le sous-espace vectoriel réel engendré par τ_z.

8. On définit sur ∂D une famille de pseudo-boules $B(z,r)$ de centre z et de rayon r en posant

$$B(z,r) = \{\zeta\in\partial D ; |\Pi_z(\zeta-z)|<r , \quad |\zeta-z|<r^2\}$$

où Π_z désigne dans $T_z(C^n)$ la projection orthogonale complexe sur $C[\chi_z]$ le sous-espace vectoriel complexe engendré par χ_z.

Soit ρ la pseudo-distance sur ∂D définie par $\rho(z,\omega)=\inf\{r$; il existe une pseudo-boule de rayon r qui contient z et $\omega\}$.

Si E est un sous-ensemble fermé de ∂D on note

$$\rho(z,E) = \inf_{\omega\in E} \rho(z,\omega).$$

9. Avec ces notations on peut reformuler un théorème de A. Davie et B. Øksendal.

10. THÉORÈME [6]. *Soit* E *un sous-ensemble fermé de* ∂D . *Si, pour tout* $\varepsilon>0$, *il existe un recouvrement de* E *par une famille de pseudo-boules* $\{B(z_i,r_i)\}_{i\in\mathbb{N}}$ *telle que l'on ait* $\sum_{i=1}^{\infty} r_i<\varepsilon$ *alors* E *est un ensemble pic pour* A(D).

La preuve utilise ce recouvrement et l'existence d'une fonction F_z qui pique en chaque point z de ∂D. On a des estimations précises sur chacune des fonctions F_{z_i} dans la pseudo-boule $B(z_i,r_i)$. On construit alors une série de fonctions holomorphes dans D et continues dans $\bar{D}\setminus E$ qui converge uniformément sur tout compact de $\bar{D}\setminus E$ et dont la somme F(z) ne s'annule que sur E et vérifie $\mathrm{Re}F(z)>0$ dans $\bar{D}\setminus E$.

11. REMARQUE. Le théorème 10 donne une condition métrique pour qu'un sous-ensemble fermé de ∂D soit un ensemble pic pour A(D). Une telle condition est vérifiée par des ensembles de structure très différente. Elle est en effet satisfaite par des sous-ensembles de ∂D très "dipersés" et aussi par des courbes de classe $C^{1+\varepsilon}$, $\varepsilon>0$, celles dont la tangente en chaque point z est contenue dans $T_z^c(\partial D)$.

12. Les sous-variété N de ∂D dont l'espace tangent est contenu dans l'espace tangent complexe ont fait l'objet d'une étude particulière. Alors que D. Burns et L. Stout [1] se posaient, pour des variétés réelles analytiques, le problème de l'interpolation réelle analytique, A. Nagel [14], G. Henkin et A. Tumanov [10] et W. Rudin [18] prouvaient avec des méthodes et des hypothèses de régularité différentes le résultat suivant.

13. THÉORÈME [18]. *Soit* N *une sous-variété de classe* C^1 *de* ∂D *tel que l'espace tangent en chaque poit* z *de* N *soit contenu dans* $T_z^c(\partial D)$; *alors tout sous ensemble compact de* N *est un ensemble pic pour* A(D).

Dans [10] et [15] on trouve une condition nécessaire pour qu'un sous-ensemble E de ∂D soit un ensemble pic pour $A(D)$.

14. THÉORÈME [15]. *Soit* E *un sous-ensemble de* ∂D *pic pour* $A(D)$ γ *une courbe sur* ∂D *de classe* $C^{1+\varepsilon}$, $\varepsilon>0$, *dont la tangente ne soit en aucun point* z *contenue dans l'espace tangent complexe* $T_z^c(\partial D)$ *et* μ *la mesure de longueur sur* γ *; alors on a*

$$\mu(E\cap\gamma) = 0.$$

Il s'en déduit donc que la condition $T_p(N)\subset T_p^c(\partial D)$ est nécessaire pour que tout sous-ensemble compact d'une sous-variété N de ∂D soit pic pour $A(D)$.

ÉTUDE DE $A^\infty(D)$.

15. Alors que le théorème 5 a son analogue dans le cas de $A^\infty(D)$ la classe des fonctions analytiques dans D, continues ainsi que toutes leurs dérivées dans $\bar{D}$, les théorème 4 n'est plus vrai pour cette classe même dans le cas $n=1$.

En effet, lorsque D est le disque unité du plan complexe, m la mesure de Lebesgue sur le cercle, les sous-ensembles parfaits de ∂D, ensembles de zéros pour $A^\infty(D)$, sont caractérisés [16],[20],[13]par la condition de Carleson

$$\int_{\partial D} \log \frac{1}{d(z,E)}\, dm(z) < \infty$$

et les seuls ensembles pics pour $A^\infty(D)$ sont les ensembles finis [20].

16. On supposera maintenant, bien que cela ne soit pas toujours nécessaire, que la frontière ∂D est de classe C^∞.

EXEMPLES D'ENSEMBLES DE ZÉROS POUR $A^\infty(D)$.

17. THÉORÈME [4]. *On note* σ *la mesure superficielle sur* ∂D. *Soit* E *un sous-ensemble fermé de* ∂D *tel que*

$$\int_{\partial D} \frac{1}{\rho(z,E)^{n-1}} \operatorname{Log} \frac{1}{\rho(z,E)}\, d\sigma(z) < \infty$$

alors E *est un ensemble de zéros pour* $A^\infty(D)$.

On utilise un recouvrement de Whitney du complémentaire de E et un procédé de série pour construire une fonction qui s'annule sur E ainsi que toutes ses dérivées.

Dans le cas du disque unité, on retrouve la condition de Carleson car la pseudo-distance se réduit alors à la distance euclidienne.

Des formes équivalentes de cette condition permettent de vérifier qu'elle est satisfaite comme celle de A. Davie et B. Øksendal dans le cas de A(D) par des sous-ensembles de ∂D très "dispersés" et les sous-ensembles compacts de courbe de classe $C^{1+\varepsilon}$, $\varepsilon>0$, dont la tangente en chaque point est dans l'espace tangent complexe [4].

EXEMPLES D'ENSEMBLES PICS ET D'ENSEMBLES D'INTERPOLATION POUR $A^{\infty}(D)$.

18. DÉFINITION. Une sous-variété M de $\mathbb{C}^n$ est totalement réelle si, pour tout point p de M, on a

$$T_p(M) \cap J\, T_p(M) = \{0\}.$$

On sait [1] que si D est un domaine strictement pseudo-convexe de $\mathbb{C}^n$ et si N est une sous-variété de ∂D telle que, en chaque point p de N, on ait

$$T_p(N) \subset T_p^c(\partial D),$$

alors N est totalement réelle.

On en déduit que, dans ce cas, on a

$$\dim_{\mathbb{R}} N \leq n-1 .$$

19. THÉORÈME [8]. *Soit* E *un sous-ensemble fermé de* ∂D *pic pour* $A^{\infty}(D)$. *Alors* E *est localement contenu dans une sous-variété* M *de* ∂D, *de dimension* n-1, *totalement réelle et telle que, en tout point* p *de* E, *l'on ait*

$$T_p(N) \subset T_p^c(\partial D).$$

On retrouve en corollaire de ce résultat la condition nécessaire sur les sous-variétés de ∂D déjà citée dans le cas de A(D). Si une sous-variété N de ∂D est telle que tout compact de N est un ensemble pic pour $A^\infty(D)$ alors on a en tout point p de N, $T_p(N)\subset T_p^c(\partial D)$. On peut remarquer également [8] que la réunion de deux sous-ensembles pics pour $A^\infty(D)$ n'est pas, en général, un ensemble pic pour $A^\infty(D)$.

Les sous-variétés de ∂D vérifiant, en tout point p de N, $T_p(N)\subset T_p^c(\partial D)$ ont fait l'objet de plusieurs travaux [14],[8],[3].

20. DÉFINITION. Un sous-ensemble E de ∂D est d'interpolation pour $A^\infty(D)$ si, pout toute fonction f de classe C^∞ sur E, il existe une fonction F de $A^\infty(D)$ telle que l'on ait $F|_E = f$.

21. THÉORÈME. *Soit* N *une sous-variété de classe* C^∞ *de* ∂D *telle que l'on ait, en tout point* p *de* N, $T_p(N)\subset T_p^c(\partial D)$ *alors tout sous-ensemble compact* K *de* N *est*

a) un ensemble de zéros pour $A^\infty(D)$,

b) un ensemble pic pour $A^\infty(D)$,

c) un ensemble d'interpolation pour $A^\infty(D)$.

En utilisant les techniques qu'il avait développées dans le cas de A(D), A. Nagel [14] a montré tout d'abord que, pour tout compact E et tout ouvert W et N vérifiant $E\subset W\subset N$, il existe un compact E_1 tel que l'on ait $E\subset E_1\subset W$ et tel que E_1 soit un ensemble de zéros pour $A^\infty(D)$. M. Hakim et N. Sibony [8] en résolvant un problème de $\bar\partial$ et un utilisant des changements de variable presque analytiques ont établi le théorème 20 a), c) et une version locale de b). Le théorème 20 b) est prouvé dans [3] où l'on utilise, outre la solution d'un problème $\bar\partial$, des techniques développées par G. Henkin et A. Tumanov dans [10].

BIBLIOGRAPHIE

[1] D. Burns, and E.L. Stout. Extending functions from submanifolds on boundary. *Duke Math. J.* 43 (1976), 391-403.

[2] J. Chaumat, et A.-M. Chollet. Ensembles de zéros, ensembles pics et d'interpolation pour A(D). *Actes du Colloque d'Anal. Harm. et Compl.* 1977. La Garde-Freinet. (Laboratoire de Math. Pures de Mar-

seille).

[3] J. Chaumat, et A.-M. Chollet. Ensembles pics pour $A^{\infty}(D)$. *Ann. Inst. Fourier* 29 (1979).

[4] A.-M. Chollet. Ensembles de zéros à la frontière de fonctions analytiques dans des domaines strictement pseudoconvexes. *Ann. Inst. Fourier* 26 (1976), 51-80.

[5] B. Cole, and R.M. Range. A- measures on complex manifolds and some applications. *J. Funct. Anal.* 11(1972), 393-400.

[6] A.M. Davie, and B.K. Øksendal. Peak interpolation sets for some algebras of analytic functions. *Pacific J. Math.* 41 (1972), 81-87.

[7] J. Détraz. Approximation et interpolation dans un domaine pseudoconvexe. *C.R. Acad. Sc. Paris* 277 (1973), 583-586.

[8] M. Hakim, et N. Sibony. Ensembles pics dans des domaines strictement pseudoconvexes. *Duke Math. J.* (1978).

[9] G.M. Henkin, and E.M. Čirka. Boundary properties of holomorphic functions of several complex variables. *J. Soviet Math.* 5 (1976), 612-687.

[10] G.M. Henkin, and A.E. Tumanov. C.R. Ecole d'été à Drogobytch (1974). (en russe).

[11] K. Hoffman. *Banach spaces of analytic functions*, Prentice Hall, New Jersey (1962).

[12] N. Kerzman, Hölder and L^p estimates for solutions of $\bar{\partial}u = f$ in strongly pseudoconvex domains. *Comm. Pure and Appl. Math.* 24(1971), 301-379.

[13] B.I. Koremblum. Functions holomorphic in a disk and smooth in its closure. *Soviet Math. Dokl.* 12(1971), 1312-1315.

[14] A. Nagel. Smooth zero sets and interpolation sets for some algebras of holomorphic functions on strictly peudoconvex domains. *Duke Math.*

J. 43 (1976), 323-348.

[15] A. Nagel, and W. Rudin, Local boundary behavior of bounded holomorphic functions. *Can. J. Math.* 30 (1978), 583-592.

[16] W.P. Novinger. Holomorphic functions with infinitely differentiable boundary values. *Illinois J. Math.* 15 (1971), 80-90.

[17] R.M. Range, Approximation to bounded holomorphic functions on stricly pseudoconvex domains. *Pac. J. Math.* 41. (1972), 203-213.

[18] W. Rudin, Peak interpolation sets of classe C^1. *Pac. J. Math.* 75 (1978), 267-279.

[19] E.L. Stout, A Rudin-Carleson theorem on balls. (non publié).

[20] B.A. Taylor, and D.L. Williams. Ideals in rings of analytic functions with smooth boundary values. *Canadian J. Math.* 22 (1970), 1266-1283.

[21] B.A. Talylor, and D.L. Williams. The peak sets of A^m. *Proc. Amer. Math. Soc.* 24 (1970), 604-605.

[22] R.E. Valskii. On measures orthogonal to analytic functions in $\mathbb{C}^n$. *Soviet Math. Dokl.* 12 (1971), 808-812.

Université de Paris-Sud

Orsay

ON CHEBYCHEV APPROXIMATION ON SEVERAL DISJOINT INTERVALS

by

W.H.J. Fuchs*

This talk is concerned with Chebychev approximation on the union K of p disjoint, bounded closed intervals I_j of the real axis ($j = 1,2...p$; $p > 1$). Let $\mathcal{P}_n$ denote the set of all polynomials of degree $\leq n$ with complex coefficients. For $f(x)$ defined on K let

$$E_n(f) = \inf_{q \in \mathcal{P}_n} ||f-g||$$

where the norm is the sup-norm on K. By a well known theorem, given $f \in C(K)$, there is a unique $p_n(x) \in \mathcal{P}_n$ such that

$$E_n(f) = ||f-p_n(x)||.$$

The asympototic behavior of $E_n(f)$ as $n \to \infty$ is determined a) by the regularity properties of f and b) by the geometry of the set K. We are mainly interested in the influence of the geometry of K on $E_n(f)$. This motivates the introduction of the classe of functions $\mathcal{R}$:

DEFINITION: $f(x) \in \mathcal{R}$ iff 1. f is defined and real-valued on K 2. On I_j f is equal to the restriction of an entire function $h_j(z)$ to I_j. 3. $f(x)$ is not the restriction of a single entire function $h(z)$ to K.

Let M be the complement of K in the completed complex plane $\mathbb{C}$ and let $k(z)$ be the Green's function of M with pole at ∞. We extend the definition of k to all of $\mathbb{C}$ by putting

$$k(x) = 0 \quad (x \in K).$$

* Research supported by grant MCS 78-00680 of the National Science Foundation.

For small $\alpha > 0$ the set

$$D_\alpha = \{z | k(z) < \alpha\}$$

will consist of p components each of which contains one and only one I_j. If $f \in \mathcal{R}$, then f has a natural holomorphic extension to D_α under these circumstances. As α increases, a value α_1, say, of α will be reached such that the boundaries of two of the components of D_α have a point in common. Unless the entire functions h_j and h_ℓ, say, which define f in these components are identical, f will not have a holomorphic extension to D_ρ with $\rho > \alpha_1$. It is not hard to see that $k(z) = \alpha_1$ contains a critical point of $k(z)$ ($\frac{\partial k}{\partial x} = \frac{\partial k}{\partial y} = 0$) at the point where two distinct components of $D_\alpha (\alpha < \alpha_1)$ begin to coalesce. In the most general case there will be a positive number β such that f(z) has a holomorphic extenstion to D_β, but not to D_ρ with $\rho > \beta$. The boundary of D_β will contain (at least) one critical point of k(z). There are p-1 critical points. They all lie on the real axis, one in each interval separating two successive I_j.

Our main results are the following.

THEOREM 1. *Let* $f \in \mathcal{R}$ *and assume that* f(z) *is holomorphic in* D_β, *but not in* D_ρ *with* $\rho > \beta$. *Then*

$$E_n(f) < An^{-\frac{1}{2}} e^{-\beta n}$$

THEOREM 2. *Under the hypotheses of Theorem 1.*

$$E_n(f) > An^{-\frac{1}{2}} e^{-\beta n} \qquad (1)$$

unless very special relations hold between the values of the $h_j(z)$ *(entering into the definition of* $f \in \mathcal{R}$) *near the critical points of* k *on* ∂D_β. *If these relations hold, the factor* $n^{-\frac{1}{2}}$ *in (1) must be replaced by an* n^{-c} *with* $c > \frac{1}{2}$.

REMARK: The letter A denotes a positive number, possibly depending of f, but independent of n. At different occurences A may have different numerical values.

THEOREM 3. *Let* $f \in \mathcal{R}$. *Denote by* $\omega(J)$ *the harmonic measure at* ∞ *of the interval* $J \subset K$ *with respect to* M. *Then*

$$(\text{\# of zeros of } f-p_n \text{ in } J)/n \to \omega(J) \quad (n \to \infty).$$

These theorems were found as (partial) answers to questions which arise in the design of electrical filters [3]. I am grateful to Dr. H. Pollak of Bell Laboratories for pointing out these questions and to Dr. J. Kaiser for most interesting information on the practical aspects of the problems.

It seems likley that the conclusion of Theorem 3 holds under the sole assumption that f(x) is holomorphic at every point of K.

H. Widom has shown [5] that the Chebychev polynomials $T_n(z)$ of K satisfy

$$\|T_n(z)\| = B_n(C(K))^n$$

where C(K) is the logarithmic capacity of K and the B_n oscillate between finite limits as $n \to \infty$. It would be interesting to know whether under the conditions of Theorem 1 the constants $\|f-p_n\| n^{\frac{1}{2}} e^{n\beta}$ also oscillate between finite limits.

SKETCH OF THE PROOF OF THEOREM 2

Let μ be a measure orthogonal to $1, x, x^2 \ldots x^n$ on K. Then

$$\left|\int_K f d\mu\right| = \left|\int_K (f-p_n) d\mu\right| \le E_n(f) \int_K |d\mu|. \tag{2}$$

This formula supplies a lower bound for $E_n(f)$. To construct μ, let $\phi(z)$ be a holomorphic function on M with a zero of order $n+2$ at ∞,

$$\phi(z) = \frac{A_1}{z^{n+2}} + \frac{A_2}{z^{n+3}} + \ldots \quad (|z| \ge B) \tag{3}$$

and suppose that $\phi(z)$ is bounded in a neighborhood of K. By Cauchy's Theorem and (3)

$$0 = \int_{|z|=B} z^j \phi(z) dz = \sum_{k=1}^{p} \int_{C_k} z^j \phi(z) dz \quad (j = 0,1,2,\ldots n)$$

where C_k is a simple closed curve surrounding I_k, but no other I_ℓ.

Contracting C_k to the interval I_k we have

$$0 = \sum_{k=1}^{p} \int_{I_k} x^j\{\phi(x-i0) - \phi(x+i0)\}dx \quad (j=0,1\ldots,n).$$

Therefore

$$d\mu = \{\phi(x+i0) - \phi(x+i0)\}dx \qquad (4)$$

defines a measure on K orthogonal to x^j $(j=0,\ldots,n)$. To find a suitable ϕ we assume ϕ in the form

$$\phi(z) = e^{\Psi(z)}$$

where

$$\mathcal{R}\Psi(z) = -(n+2)k(z) + \sum_{k=1}^{p} \lambda_k\omega_k(z).$$

Here $\omega_k(z)$ is the harmonic measure at z of I_k with respect to M. The λ_j are real constants which must be adjusted so that the conjugate function of $\mathcal{R}\Psi$ is single-valued in M. Such λ_j satisfying the condition $|\lambda_j|<A$ can always be found [5]. To calculate $\int_K fd\mu$, where μ is given by (4), we use again Cauchy's Theorem and write

$$\int_K fd\mu = \int_{\partial D_\beta} f(z)\phi(z)dz .$$

The contour ∂D_β passes through one (or more) critical points of k(z). Draw lines perpendicular to the real axis at these points and replace the integration around D_β by an integration around the boundaries of the regions into which these lines divide $D_{\beta+\varepsilon}$ (f(z) has a unique analytic continuation into each of these regions, if the positive number ε is sufficiently small). It is not hard to see that for large n the main contribution to $\int fd\mu$ comes from the immediate neighborhood of the critical points. Evaluating these contributions by Laplace's method yields the theorem.

Theorem 1 is proved by contructing an approximating polynomial to f by interpolation.

The points of interpolation are chosen at the zeros of the nth generalized Faber polynomial q(z) introduced in [5]. This polynomial is given by

$$q_n(z) = \frac{1}{2\pi i} \int_{|\zeta|=R} \frac{\phi_1(\zeta)d\zeta}{\zeta - z} \qquad \text{(R large)}$$

where $\phi_1(\zeta)$ is a function similar to the function $\phi(\zeta)$ used above, but with a pole of order $n+1$ at ∞, instead of a zero of order $n+2$.

SKETCH OF THE PROOF OF THEOREM 3

We shall consider the special case $J = I_j$ and we shall write ω_j for $\omega(I_j)$. The general case is then proved by a fairly routine elaboration of the arguments in this special case. Let $Z(S)$ denote the number of zeros of $f-p_n$ in the point set S. It will be enough to prove: Given $\eta > 0$, $\exists n_0(\eta)$ such that

$$Z(I_j) < \omega_j n + \eta n \quad (\omega_j = \omega(I_j), n > n_0(\eta)) \quad . \tag{5}$$

For, by a well known characterisation of polynomials of best approximation [1, p.75] there are $n+2$ points of K,

$$y_1 < y_2 \ldots < y_{n+2}$$

such that $r(y) = f(y) - p_n(y)$ has different signs at $y=y_j$ and $y=y_{j+1}$ $(j = 1,2\ldots,n+1)$. Since only $p-1$ of the intervals (y_j, y_{j+1}) can contain points not in K, we have therefore $Z(K) \geq n+1-(p-1) = n+2-p$. Hence, if (5) holds,

$$\begin{aligned} Z(\omega_j) &= Z(K) - \sum_{k \neq j} Z(\omega_k) \\ &\geq n+2-p- \sum_{k \neq j} \omega_k n - (p-1)\eta n \qquad (n > n_0(\eta)) \qquad (6) \\ &\geq \omega_j n - p - (p-1)\eta n \ , \end{aligned}$$

since $\sum_{k=1}^{p} \omega_k = 1$. Our assertion follows now from (5) and (6).

By a well known formula for the harmonic measure [2,p.30] and by Gauss' Theorem

$$\begin{aligned} \omega_j &= \frac{1}{2\pi} \int_{I_j} \{\frac{\partial h}{\partial y}(x+i0) - \frac{\partial k}{\partial y}(x-i0)\}dx \qquad (7) \\ &= \frac{1}{2\pi} \int_C \frac{\partial k}{\partial n}(\zeta)\,|d\zeta| \end{aligned}$$

where C is a contour encircling I_j once, but no other I_k. Choose α so small ($\alpha<\alpha_0$) that there is a component E_α of D_α containing I_j, but no other I_k. Choose $C=C_\alpha=\partial E_\alpha$. By the argument principle and (7)

$$Z(I_j)-\omega_j n\le Z(E_\alpha)-\omega_j n = \frac{1}{2\pi}\int_{C_\alpha} \frac{\partial}{\partial n}\ \ell(\zeta)\,|d\zeta| = \Psi(\alpha) \qquad (8)$$

say, where

$$\ell(\zeta) = \log|f(\zeta)-p_n(\zeta)| + n\beta - nk(\zeta).$$

(It is assumed that $f-p_n\neq 0$ on C_α.) By a geometrical argument

$$\Psi(\alpha) = \frac{d}{d\alpha}\ \xi(\alpha)$$

where

$$\xi(\alpha) = \frac{1}{2\pi}\int_{C_\alpha} \ell(\zeta)\,\frac{\partial k}{\partial n}(\zeta)\ |d\zeta| \qquad (9)$$

The functions $\xi(\alpha)$ is a continuous function of α where it is defined, so that by integration over successive ranges in which $f-p_n\neq 0$ on C_α,

$$\xi(\alpha_2) - \xi(\alpha_1) = \int_{\alpha_1}^{\alpha_1} \Psi(\alpha)\,d\alpha. \quad (D<\alpha_1<\alpha_2<\alpha_0) \qquad (10)$$

By classical results of J.L. Walsh [4, p.79] on the maximal convergence of polynomials of best approximation

$$\ell(\zeta) < \eta n \qquad (\zeta\in D_\beta) \qquad (11)$$

and therefore, by (9) and (7),

$$\xi(\alpha) < \eta n\omega_j < \eta n \qquad (\alpha<\alpha_0). \qquad (12)$$

If we can show that there is a γ such that

$$\xi(\eta) > A\ \eta n \qquad (\eta < \gamma < \alpha_0), \qquad (13)$$

then it will follow from (12), (13) and (10) that the increasing function $\Psi(\alpha)$ satisfies

$$\Psi(\alpha) < A\eta n$$

for some α in $0 < \alpha < \alpha_0$. This implies (5) [with η replaced by $A\eta$] because of (8). It remains to prove (13). The proof is based on two facts:

1. Given $\delta>0$, $\exists\ n_0(\delta)$ such that for $n>n_0(\delta)$ every interval $J\subset K$ of length δ contains a point x (depending on n) such that $\ell(x)>-\eta n$.

2. For $\zeta\in C_\eta$ sufficiently close to $x\in I_j$

$$0<\frac{\partial k}{\partial n}(\zeta)<-A\eta\frac{\partial g}{\partial n}(\zeta,x,E_\eta), \tag{14}$$

where $g(z,x,E_\eta)$ is the Green's function of E_η with pole at x.

The proof of 1. makes heavy use of the fact that $f\in\mathcal{R}$, so that the conclusions of Theorems 1 and 2 hold.

The proof of 2. is based on elementary estimates of $g(z,x,E_2)$ from below.

Once 1. and 2. are established one can show that C_η can be divided into $0(1/\eta)$ arcs Γ_ν so that (14) applies on Γ_ν with an x for wich $\ell(x)>-\eta n$. Now

$$\int_{\Gamma_\nu}\ell(\zeta)\frac{\partial k}{\partial n}(\zeta)|d\zeta|>\int_{\Gamma_\nu}(\ell(\zeta)-\eta n)\frac{\partial k}{\partial n}(\zeta)|d\zeta|.$$

By (11) and (14)

$$\int_{\Gamma_\nu}\ell(\zeta)\frac{\partial k}{\partial n}(\zeta)|d\zeta|>-A\eta\int_{\Gamma_\nu}(\ell(\zeta)-\eta n)\frac{\partial g}{\partial n}(f,x,E_\eta)\ |d\zeta| \tag{15}$$

$$>-A\eta\int_{C_\eta}(\ell(\zeta)-\eta n)\frac{\partial g}{\partial n}(\zeta,x,E_\eta)|d\zeta|$$

On C_η, $\ell(\zeta)-\eta n$ is equal to the function

$$\log|f(\zeta)-p_n(\zeta)|+\beta_n-2\eta_n=\phi(\zeta)$$

which is subharmonic in E_η. The last integral in (15) is 2π times the boundary harmonic majorant of $\phi(\zeta)$ in E_η evaluated at x. Therefore

$$\int_{\Gamma_\nu}\ell(\zeta)\frac{\partial k}{\partial n}(\zeta)|d\zeta|>-2\pi A\eta(\ell(x)-\eta n)=2\pi A\eta^2 n\ .$$

Summing over the $0(1/\eta)$ different values of we obtain (13) and the proof is completed.

REFERENCES

1 E. W. Cheney, *Introduction to Approximation Theory*, McGraw Hill, 1966.

2 R.Nevanlinna, *Eindeutige Analytische Funktionen*, 2nd edition, Springer 1953.

3 L. R. Rabiner and C.M. Rader (editors), *Digital Signal Processing*, IEEE Press, 1972.

4 J. L. Walsh, *Interpolation and Approximation by Rational Functions in the complex Domain*, 2nd ed. Am. Math. Soc. Colloquium Publications, Vol. 20, 1935.

5 H. Widom, Extremal Polynomials Associated with a System of Curves in the Complex Plane, *Advances of Math*. 3 (1969), 127-232.

Cornell University
Ithaca.

APPROXIMATION IN L^p BY ANALYTIC AND HARMONIC FUNCTIONS

by

Lars Inge Hedberg

The results reported on in the talk have appeared in [1]. A summary of some of the results in that paper will appear in [2], and for this reason only a short summary is given here.

Let K be a compact set in $\mathbb{R}^d$, let $L^p(K)$ denote L^p with respect to d-dimensional Lebesgue measure restricted to K, and let $L_h^p(K)$ be the subspace of $L^p(K)$ consisting of functions harmonic in the interior of K. The problem is to characterize those sets K such that the functions harmonic on K (i.e., on some neighborhood of K) are dense in $L_h^p(K)$. Except in the case when K has no interior this problem turns out to be harder than the corresponding problem for analytic functions (and $K \subset \mathbb{C}$), which was solved earlier by V.P. Havin, T. Bagby, and the author. The complication comes from the fact that the Sobolev space $W^{1,q}$ is closed under truncation, but $W^{2,q}$ is not.

In [1] it is shown that the approximation problem is related to a "spectral synthesis" problem in Sobolev spaces. This problem is given a partial solution, and as a consequence sufficient (but not necessary) conditions on K to kave the above approximation property are obtained.

REFERENCES

[1] L. I. Hedberg, Two approximation problems in function spaces, *Ark. Mat.* 16(1978), 51-81.

[2] L. I. Hedberg, Approximation in L^p by analytic and harmonic func-

tions, *Proc. Symp. Pure Math.* (1978 Summer Institute on Harmonic Analysis in Euclidiean Spaces and Related Topics), Amer. Math. Soc., Providence, R.I., 1979.

University of Stockholm, Sweden.

GAP-INTERPOLATION THEOREMS FOR ENTIRE FUNCTIONS

by

Nigel Kalton

Lee Rubel*

We give a preliminary report here on our work on the problem of interpolation on a sequence $Z = (z_n)$ of non-zero complex numbers by entire functions f of the form

$$f(z) = \sum_{\lambda \in \Lambda} a_\lambda z^\lambda , \tag{1}$$

where Λ is a given sequence of positive intergers. This problem seems first to have been raised in [2]. In this summary, we limit ourselves to stating the main theorem (and two of its immediate corollaries), which gives a necessary and sufficient condition that such interpolation be possible. This condition is not easy to work with, and we are in the process of finding some more tractable conditions that either imply it or are implied by it. They involve, somewhat to our surprise, questions of diophantine approximation and p-adic analysis. The proof of our theorem, which we omit, involves functional analysis, specifically the main result of [1]. We expect to publish the details elsewhere.

In what follows, Z is a sequence (z_n), (perhaps a terminating one) of non-zero complex numbers, and Λ is a sequence of positive integers. We suppose that $|z_{n+1}| \geq |z_n|$ for all n.

* The research of the authors was supported by separate grants from the National Science Foundation.

DEFINITION. Λ is an interpolating sequence of exponents for Z if for every sequence (w_n) of complex numbers, there exists an entire function f of the from (1) such that $f(z_n) = w_n$ for every n.

DEFINITION. Λ is Z linearly independent if

$$\sum_{n=1}^{N} a_n z_n^{\lambda} = 0 \quad \text{for all } \lambda \in \Lambda$$

implies that each a_n vanishes.

DEFINITION. Λ is asymptotically Z linearly independent if for every $\rho > 0$, there exists an $N(\rho)$ so that

$$\left|\sum_{n=1}^{K} a_n z_n^{\lambda}\right| \leq \rho^{\lambda} \quad \text{for all } \lambda \in \Lambda$$

implies that $a_n = 0$ for all $n \geq N(\rho)$.

DEFINITION. Λ is totally Z linearly independent if it is both Z linearly independent and asymptotically Z linearly independent.

THEOREM. *Λ is an interpolating sequence of exponents for Z if and only if Λ is totally Z linearly independent.*

We state two simple consequences of this theorem.

COROLLARY 1. *Λ is an interpolating sequence of exponents for every sequence Z that has no finite limit point and has $|z_i| \neq |z_j|$ for $i \neq j$, if and only if Λ is infinite.*

COROLLARY 2. *Λ is an interpolating sequence of exponents for every terminating (i.e. finite) sequence Z if and only if every arithmetic progression $\{an+b\}$, $a \neq 0$, contains at least one element of Λ.*

Corollary 1 was proved by other means in [2]. Corollary 2 is an immediate consequence of our main theorem and the Skolem-Lech theorem [3]. We are grateful to D.J. Newman for pointing out to us the relevance of the Skolem-Lech theorem, an interesting feature of which is that the only known proof seems to be dependent on p-adic analysis.

REFERENCES

[1] Paul M. Gauthier and Lee A. Rubel, Interpolation in separable Fréchet

spaces with applications to spaces of analytic functions, *Canad. J. Math.* 27 (1975), 1110-1113.

[2] Nigel Kalton, On summability domains, *Proc. Camb. Philo. Soc.* 73 (1973), 327-338.

[3] Christer Lech, A note on recurring series, *Arkiv för Mat.* 2 (1952), 417-421.

University College, Swansea

University of Illinois at Urbana-Champaign

UNIFORM APPROXIMATION ON CERTAIN UNBOUNDED SETS IN $\mathbb{C}^n$

by

Edgar Lee Stout*

I. The work reported on here is motivated by a theorem of Carleman [1] to the effect that *given a continuous function* $f: \mathbb{R} \to \mathbb{C}$ *and a continuous, positive function* ε *on* $\mathbb{R}$, *there exists an entire function* F *of one complexe variable with* $|F(x)-f(x)| < \varepsilon(x)$ *for all* $x \in \mathbb{R}$: Not only do we have uniform approximation but the approximation can be made to improve arbitrarily rapidly at infinity. Hoishen [3] and Scheinberg [5] showed that the same kind of asymptotically improving approximation can be obtained of $\mathbb{R}^n$ with entire functions of n complex variables, and Hoischen noted that it is possible to approximate not only f but also finitely many of its (partial) derivatives à la Carleman. A subsequent result of Numemacher [4] is that *if* $\Sigma \subset \mathbb{C}^n$ *is a totally real* C^1 *(closed) submanifold, and if* $f: \Sigma \to \mathbb{C}$ *is a continuous function and if* ε *is a positive continuous function, then there is* F *holomorphic on a neighborhood of* Σ *with* $|f-F| < \varepsilon$ *on* Σ. In this result, the neighborhood apparently depends on the particular function to be approximated, and elementary convexity considerations show that in general the approximation cannot be made by entire functions.

II. By using the techniques of complexification and embedding, it is possible to derive from Hoishen's theorem a stronger approximation theorem on real analytic totally real manifolds.

---

*The research reported on in this abstract was supported by the National Science Fondation through grants MCS 78-02139 and MCS 76-06325.

THEOREM 1. *Let Σ be a real analytic, totally real submanifold of a complex manifold M. There exists a neighborhood Ω of Σ in M such that given $f \in C^p(\Sigma)$, $1 \leq p < \infty$, given a positive continuous function ε on Σ, given s vector fields $\xi_1, \ldots, \xi_s$ of class C^p on Σ, and given a finite collection $P_1, \ldots, P_r$ of polynomials in s noncommuting indeterminants, each P_j of degree no more than p, there exists F holomorphic in Ω with*

$$|P_j(\xi_1, \ldots, \xi_s)\,(f-F)| < \varepsilon$$

on Σ.

This theorem is of interest in its own right, but it is also of interest because it suggests very strongly what may be true on less smooth manifolds Σ.

III. We also have an approximation theorem on certain Lipschitz graphs. Consider $\mathbb{R}^m \subset \mathbb{C}^m$ and in $\mathbb{R}^m$ consider the real subspace $\mathbb{R}^p = \{(x_1, \ldots, s_p, 0, \ldots, 0) : x_1, \ldots, x_p \in \mathbb{R}\}$. Let $\psi_{p+1}, \ldots \psi_m$, $\phi_1, \ldots, \phi_m$ be continuous real valued functions on $\mathbb{R}^p$. For $x \in \mathbb{R}^p$, put $\psi(x) = (\psi_{p+1}(x), \ldots, \psi_m(x))$ and $\phi(x) = (\phi_1(x), \ldots \phi_m(x))$. If we then set $\Psi(x) = (x, \psi(x)) \in \mathbb{R}^m$, we can speak of the point $\Psi(x) + i\phi(x)$. Define Σ to be the manifold given by

$$\Sigma = \{\Psi(x) + i\phi(x) \;:\; x \in \mathbb{R}^p\} \subset \mathbb{C}^m$$

THEOREM 2. *Assume the functions ψ_j, $p+1 \leq j \leq m$ and ϕ_j, $1 \leq j \leq m$, to satisfy a Lipschitz condition with Lipschitz constant k such that*

$$\tan^{-1} \frac{2k}{1-k^2} < \frac{\pi}{2p} .$$

If f is a bounded, uniformly continuous function on Σ and if is a positive real number, then there is an entire function F on $\mathbb{C}^n$ such that $|F(z) - f(z)| < \varepsilon$ for all $z \in \Sigma$.

N.B. In this statement, ε is a constant, not a function as in the preceding paragraphs.

Notice also that in Theorem 2, Σ is not assumed explicitly to be totally real. By Rademacher's theorem, Σ has a tangent plan at almost every point, and it follows from work of Ferrier and Siddiqi [2] that these tangent planes are totally real.

The details of these results will appear in due course.

REFERENCES

[1] T. Carleman, Sur un théorème de Weierstrass, *Arkiv för Mat. Ast. Fys.* 20B(1927), 1-5.

[2] J.-P. Ferrier and J. Siddiqi, Convexité polynomiale et approximation pondérée sur certaines sous-variétés réelles de $\mathbb{C}^n$, *J. Math. Pures et App.* 55(1976), 445-466.

[3] L. Hoischen, Eine Verschärtung eines Approximationssatz von Carleman, *J. Approximation Theory*, 9(1973), 272-277.

[4] J. Nunemacher, Approximation theory on totally real submanifolds, *Math. Ann.* 224 (1976), 119-141.

[5] S. Scheinberg, Uniform approximation by entire functions, *J. Anal. Mathématiques* 24 (1976), 16-18.

University of Washington

Seattle

UNIFORM APPROXIMATION ON SMOOTH POLYNOMIALLY CONVEX SETS

by

Barnet M. Weinstock

1. INTRODUCTION. The work reported on in this lecture is based on a result of Wermer [12] who studied the following problem:

Let X be a closed disc in $\mathbb{C}$ and suppose that f is a complex-valued function on X of class C^1. Let $[z,f]_X$ denote the uniform closure of the polynomials in z and f. Find conditions which imply that $[z,f]_X = C(X)$, or more generally describe $[z,f]_X$.

A necessary condition for $[z,f]_X = C(X)$ is that every complex homomorphism of the algebra $[z,f]_X$ be evaluation at some point of X, or equivalently, that if $\tilde{X}$ is the graph of f, then $\tilde{X}$ is a polynomially convex subset of $\mathbb{C}^2$.

THEOREM (Wermer [12]): *Assume $\tilde{X}$ is polynomially convex.*

(i) If $\partial f/\partial \bar{z}$ is nowhere zero on X then $[z,f]_X = C(X)$.

(ii) Let $E = \{z : \partial f/\partial \bar{z} = 0\}$. Then $[z,f]_X = \{g \in C(X) : g|_E \in R(E)\}$, where R(E) is the uniform closure on E of the space of rational functions with poles off E.

Wermer's result was generalized by Freeman [2] in two directions. He considered, in place of a disc, a two-dimensional compact (real) manifold M with boundary, and, in place of z and f, a separating family F of complex-valued functions on M containing the constants. He introduced the set

$$E = \{X \in M : df_1 \wedge df_2(X) = 0 \text{ for all } f_1, f_2 \in F\}$$

to study the algebra A of uniform limits of polynomials in F.

THEOREM (Freeman [2]): *Assume that the complex homomorphisms of*

A are given by evaluations at the points of M.

(i) If E is empty then $A = C(M)$

(ii) In general, $A=\{g\varepsilon C(M):g|_E \ \varepsilon A|_E\}$

Somewhat later, in [3], Freeman extended his theorem to n-dimensional manifolds M under the additional hypothesis that M and the functions in F are real analytic.

This problem was reformulated in more geometric language by various authors (cf.[10] ,[6],[5].) In the notation of Wermer's theorem, $\tilde{X}$ is a real submanifold of $\mathbb{C}^2$, and the condition that $\partial f/\partial\bar{z}$ doesn't vanish on X is equivalent to the condition that this submanifold has no complex tangents, i.e., that at no point is the (real) tangent plane a complex line. The term "totally real" was introduced to refer to such manifolds and their higher-dimensional generalizations -- manifolds with no complex line in the tangent space at any point. Part (i) of Wermer's (and Freeman's) theorem was recast as follows:

If M is a totally real, smooth submanifold of $\mathbb{C}^n$ and K is a compact subset of M, then every continuous function on K is a uniform limit of functions holomorphic on a neighborhood of K. If, in addition, K is polynomially convex, then every continuous function on K is the uniform limit on K of polynomials.

This was proved by Hörmander and Wermer [5] under the assumption that "smooth" means "class C^k with $k>(\dim M)/2$" and by Harvey and Wells [4] for M of class C^1. Of course, in either case, the statement concerning approximation by polynomials follows immediately from the Oka-Weil theorem. (Wermer's book [13] is an excellent reference for the Oka-Weil theorem and the general subject of uniform approximation in several complex variables.)

As regards part (ii) of Wermer's (and Freeman's) theorem, that is, in the presence of an exceptional set, results were obtained by Hörmander and Wermer [5] and by Fornaess [1], who used a construction due to Hörmander and Wermer. Indeed Fornaess succeeded in generalizing part (ii) of Freeman's theorem from M and F real-analytic to M and F of class C^k, with k as in Hörmander-Wermer.

2. STATEMENT OF RESULTS. The purpose of this lecture is to present the following new result for the C^1 case:

THEOREM: *Let X be a compact set in* $\mathbb{C}^n$, *N a neighborhood of X,* $f_1,\ldots,f_n \varepsilon C^1(N)$. *Let A be the uniform closure on X of the polynomials in* $z_1,\ldots,z_n$, $f_1,\ldots,f_n$. *Let* $\tilde{X} = \{(z,f(z)) \in \mathbb{C}^{2n}: z \varepsilon X\}$. *Assume that* $\tilde{X}$ *is polynomially convex. Let E be the subset of X where* $\det(\partial f_i/\partial \bar{z}_j) = 0$. *Then* $A = \{g \varepsilon C(X): g|_E \varepsilon A|_E\}$.

The technique of the proof is a generalization of Wermer's original proof in [12]. The idea is to replace Wermer's use of the Cauchy integral formula with a suitable Cauchy-Fantappie kernel, in somewhat the same manner as in the author's proof [8] of the local version [11] of part (i) of Wermer's theorem above.

Here is the basic outline: It suffices to show that if μ is a complex Borel measure on X such that $\int f\, d\mu = 0$ for all $f \varepsilon A$ then $\mu = 0$ on X-E. This will follow if each point z of X-E has a neighborhood U in $\mathbb{C}^n$ such that

$$\int \phi(z)\, d\mu(z) = 0 \quad \text{for all} \quad \phi \varepsilon C_0^\infty(U).$$

To achieve this we represent ϕ as

$$\phi(z) = \int_U \Omega(\zeta,z) \wedge \bar{\partial}\phi(\zeta) \qquad z \varepsilon N$$

where Ω is a Cauchy-Fantappie form with the property that for each $\zeta \varepsilon U$ and each coefficient K_j of Ω, $\int_X K_j(\zeta,z)\, d\mu(z) = 0$. The desired result then follows from Fubini's theorem.

3. CONSTRUCTION OF THE CAUCHY-FANTAPPIE FORM: We recall that Cauchy-Fantappie forms on an open set N in $\mathbb{C}^n$ are constructed as follows: (Cf. [8] and the references there.) Let G_j, $j = 1,\ldots,n$ be functions of class C^1 on U×N. Let $G(\zeta,z) = \sum(\zeta_j - z_j)G_j(\zeta,z)$. Assume that G vanishes only when $\zeta = z$. Define $\Omega(\zeta,z)$ by

$$\Omega(\zeta,z) = (n-1)!/(2\pi i)^n \sum (-1)^{j-1} G_j/G^n\ \bar{\partial}G_1 \wedge \ldots \wedge \bar{\partial}G_j \wedge \ldots \wedge \bar{\partial}G_n \wedge d\zeta_1 \wedge \ldots \wedge d\zeta_n.$$

Then $\Omega(\zeta,z)$ is a Cauchy-Fantappie form in N.

If we assume further that for each j, $G_j(.,z)G^{-n}(.,z)\in L^1_{\ell oc}$ uniformly for z in compact subsets of N then we obtain the representation formula for $\phi\in C_0^\infty(N)$ stated above. Indeed this representation formula is a simple consequence of the following basic facts about Cauchy-Fantappie forms: Let D be a smoothly bounded subdomain of N. Then

(a) $\int_{\partial D} \Omega(\zeta,z) = 1$ for each z D, and

(b) $\bar\partial_\zeta \Omega = 0$ if $\zeta\in D-\{z\}$.

The construction of a Cauchy-Fantappie form with the orthogonality property mentionned at the end of the previous section is performed in two stages. We obtain, for a given point of X-E, a neighborhood U for which a suitable function G can be defined on UxU directly in terms of the given functions $f_1,\dots,f_n$. Next we obtain a modified G, defined on UxN now, by using the polynomial convexity of $\tilde X$ and solving a Cousin problem, much as in Wermer's proof [12], except that we must use a result from [7] which states that since the Cousin data depend smoothly on $\zeta\in U$, we can find a solution to the Cousin problem which also depends smoothly on $\zeta\in U$.

LEMMA 1. *Let $a\in X-E$. Let $F(a)$ [resp. $\bar F(a)$] denote the matrix of z-derivatives [resp. $\bar z$-derivatives] of the mapping f. Define $g(\zeta,z,w)$ by*

$$g(\zeta,z,w) = -(\zeta-z).\bar F(a)^{-1}\,[f(\zeta)-w-F(a)(\zeta-z)]$$

where $z.w$ denotes the standard bilinear form on $\mathbb{C}^n$. Then there is a neighborhood U of a such that if $\zeta,z\in U$ then

$$g(\zeta,z,f(z)) = -|z-\zeta|^2 + q(z,\zeta)$$

where $|q(z,\zeta)| <\cdot |z-\zeta|^2$ if $z \neq \zeta$.

LEMMA 2. *Let U be as in Lemma 1. There is a neighborhood W of $\tilde X$ in $\mathbb{C}^{2n}$ and a function $H(\zeta,z,w)$ with the following properties:*

(a) H is holomorphic as a function of (z,w) in W for each fixed $\zeta\in U$;

(b) H is of class C^1 in ζ;

(c) there is a function $k(\zeta,z,w)$*, defined on* $U\times U\times U$*,* C^1 *in* ζ *, holomorphic in* z,w*, and non-vanishing such that* $H = kg$*, where* g *is as in Lemma 1;*

(d) $H(\zeta,z,w) \neq 0$ *if* $(z,w) \notin U\times U$*;*

(e) there is a circular sector S *centered at* 0 *and symmetric about the positive* x*-axis such that* $H(\zeta,z,f(z)) \notin S-\{0\}$ *if* $z\in X$, $\zeta\in U$.

The proof of Lemma 2 is similar to that of Lemma 14.2 in Wemer's book [13] with Proposition 2 of [7] used to insure smooth dependence of H on ζ.

LEMMA 3. *Define* $G(\zeta,z)$ *on* $U\times N$ *by* $G(\zeta,z)=H(\zeta,z,f(z))$. *There exist* $G_1(\zeta,z),\ldots,G_n(\zeta,z)$*, of class* C^1 *in* ζ *for fixed* z *in* N *such that* $G(\zeta,z)=\sum(\zeta_j-z_j)G_j(\zeta,z)$. *Moreover, for each fixed* $\zeta\varepsilon U$*, each* G_j *is in the algebra* A*, and also* $|G_j|< C|z-\zeta|$.

To prove Lemma 3 we use Proposition 5 of [7] to write $H(\zeta,z,w)$ as

$$H(\zeta,z,w)=\sum(\zeta_j-z_j)H_j(\zeta,z,w)$$

where each H_j is continuously differentiable in ζ and holomorphic in z and w. Since $\tilde{X}$ is polynomially convex, for each fixed ζ in U, each function H_j is the uniform limit on $\tilde{X}$ of polynomials in z and w. If we define $G_j(\zeta,z)$ as $H_j(\zeta,z,f(z))$ then all the conditions of the Lemma are satisfied, since the estimate follows easily from the Taylor expansion of G in view of (c) of Lemma 2 and the definition of the function g.

The Cauchy-Fantappie form constructed from the function G as in the beginning of this section will have the desired orthogonality property. The factors involving the G_j's and their $\bar{\zeta}$-derivatives are in A for each fixed ζ in U, while the denominator G^{-n} is the dominated pointwise limit on $N-\{\zeta\}$ of a sequence of elements of A. This follows from Lemma 2(e) as in Lemma 14.3 of [13] and [8, p. 62]. Notice that to demonstrate the orthogonality we can confine our attention to G on VxN where V is a neighborhood of the point a in X-E where we may assume that the cut-off function of Lemma 4 is identically one.

4. CONCLUDING REMARKS. When the set E in the statement of the theorem is empty, the graph of the mapping f is a totally real submanifold of $\mathbb{C}^{2n}$. The method outlined above thus gives a new proof for this case (polynomially convex graph) of the theorem of Harvey and Wells [4], one which does not require the construction of strongly pseudoconvex neighborhoods or the derivation of uniform estimates for the inhomogeneous Cauchy-Riemann equations. It would be desireable to extend this method to prove the approximation theorem for polynomially convex totally real submanifolds generally. The main difficulty which one would encounter is the passage from local approximation to global approximation. The "hyperfunction technique" of Harvey and Wells seems to require a sort of "super-local" approximation, which is an easy consequence of their approach via the solution of an inhomogeneous Cauchy-Riemann equation. It is not clear whether the present method using general Cauchy-Fantappie forms will be adequate.

When E is not empty, our theorem reduces approximation on $\tilde{X}$ by polynomials to approximation on $\tilde{E}$ by polynomials, where $\tilde{E}$ is the graph of $f|E$. Thus, if $\tilde{E}$ lies on a totally real submanifold of $\mathbb{C}^{2n}$ then every continuous function on $\tilde{X}$ is the uniform limit on $\tilde{X}$ of polynomials. (This follows from the fact that $\tilde{E}$ is polynomially convex). A natural question to ask now is: When will $\tilde{E}$ have this property? The author hopes to treat this question in subsequent work.

Complete details of the proof outlined above will appear elsewhere.

REFERENCES

[1] J. E. Fornaess, Uniform approximation on manifolds, *Math. Scand.*, 31 (1972) 166-170.

[2] M. Freeman. Somme conditions for uniform approximation on a manifold. *Function Algebras*, Scott, Foresman, Glenview, Ill. (1965).

[3] M. Freeman, Uniform approximation on a real-analytic manifold, *Trans. Amer. Math. Soc.* 143 (1969), 545-553.

[4] R. Harvey and R. O. Wells, Jr., Holomorphic approximation and hyperfunction theory on a C^1 totally real submanifold of a complex manifold, *Math. Ann.* 197 (1972), 287-318.

[5] L. Hörmander and J. Wermer, Uniform approximation on compact subsets in $\mathbb{C}^n$, *Math. Scand.* 23 (1968), 5-2.

[6] R. Nirenberg and R. O. Wells, Jr., Approximation theorems on a differentiable submanifold of a complex manifold, *Trans. Amer. Math. Soc.* 142 (1969), 15-35.

[7] B. Weinstock, Inhomogeneous Cauchy-Riemann systems with smooth dependence on parameters, *Duke Math. J.* 40 (1973), 307-312.

[8] B. Weinstock, A new proof of a theorem of Hörmander and Wermer, *Math. Ann.* 200(1976), 59-64.

[9] B. Weinstock, Uniform approximation and the Cauchy-Fantappie integral. *Proc. Symp. Pure Math.* 30 (1977), 187-191.

[10] R. O. Wells, Jr., Holomorphic approximation on a real-analytic submanifold of a complex manifold. *Proc. Amer. Math. Soc.* 17 (1965), 1272-1275.

[11] J. Wermer, Approximation on a disc. *Math. Ann.* 155 (1964), 331-333.

[12] J. Wermer. Polynomially convex discs. *Math. Ann.* 158 (1965), 6-10.

[13] J. Wermer. *Banach Algebras and Several Complex Variables*. Springer-Verlag, New York, 1976.

Department of Mathematics
University of North Carolina at Charlotte
Charlotte, NC 28223.

SUPPLEMENTARY TALKS

L'APPROXIMATION ENTIÈRE SUR LES ARCS ALLANT A L'INFINI DANS $\mathbb{C}^n$

par

Bernard Aupetit*

INTRODUCTION. En 1927, T. Carleman [4] a généralisé le théorème d'approximation de Weierstrass en prouvant que toute fonction continue sur la droite réelle est limite uniforme de fonctions entières. Il mentionne également que le même argument s'applique à un arc de Jordan localement rectifiable allant à l'infini dans le plan complexe. En 1939, M. Keldych et M. Lavrentieff [6] ont caractérisé les fermés connexes sur lesquels toute fonction continue peut être approximée asymptotiquement par des fonctions entières, ce qui s'applique au cas des courbes de Jordan localement rectifiable allant à l'infini. Tout cela est un cas particulier du théorème d'Arakélian [3,5] qui affirme que pour tout fermé F du plan complexe, ne séparant pas le plan et localement connexe à l'infini, toute fonction continue sur F et holomorphe à l'intérieur de F est limite uniforme de fonctions entières. Le théorème d'Arakélian, qui est une généralisation du théorème de Mergelyan, ne peut évidemment pas se généraliser dans $\mathbb{C}^n$, avec des conditions topologiques. En 1976, J. Nunemacher[7,8] a pu obtenir, en utilisant les représentations intégrales de G.M. Henkin, le résultat suivant: si V est une variété de type C^1, connexe et totalement réelle de $\mathbb{C}^n$, alors toute fonction continue sur V peut être approximée uniformément par des fonctions holomorphes dans un voisinage de V, voisinage qui dépend de la

* Travail subventionné par le Conseil national de recherches en sciences naturelles et en génie du Canada (A 7668) et le Ministère de l'Education du Québec (Subvention FCAC).

fonction donnée. Récemment E.L. Stout a obtenu dans certains cas des résultats globaux où le voisinage ne dépend pas de la fonction continue donnée mais seulement de la variété (voir texte plus haut [12]).

Tout cela nous amène naturellement à la question suivante: Si Γ est un arc de Jordan allant à l'infini dans $\mathbb{C}^n$ - c'est-à-dire que Γ admet un paramétrage γ de $\mathbb{R}$ dans $\mathbb{C}^n$ tel que $\|\gamma(t)\|$ tend vers l'infini quand $|t|$ tend vers l'infini - est-ce que toute fonction continue sur Γ est limite uniforme de fonctions entières? Sous cette forme ce problème est beaucoup trop général pour être vrai, car les exemples de J. Wermer et A.G. Vitushkin([13], p.81 et p.146)montrent que l'approximation polynomiale n'a même pas lieu sur certains arcs de Jordan compacts de $\mathbb{C}^n$, pour $n\geq 2$. Par contre on sait que si les arcs de Jordan compacts sont réguliers - c'est-à-dire de type C^1 morceaux - il y a approximation polynomiale (E. Bishop, G. Stolzenberg [10]) et plus généralement si les arcs sont rectifiables (H. Alexander [1]). Ce sont toutes les remarques qui précèdent qui nous ont amené à conjecturer le resultat suivant:

THÉORÈME 1. *Soient Γ un arc de Jordan régulier allant à l'infini dans $\mathbb{C}^n$ et ε une fonction continue strictement positive sur Γ, alors quelle que soit la fonction continue f sur Γ il existe g entière sur $\mathbb{C}^n$ telle que $|f(z)-g(z)|<\varepsilon(z)$, pour tout z de Γ.*

E.L. Stout et moi-même avions résolu partiellement cette conjecture par des méthodes différentes. Après plusieurs discussion, lors de la conférence sur l'approximation complexe, tenue à Québec de 3 au 8 juillet, en suivant ma méthode, H. Alexander a finalement réussi à prouver le lemme 3 sur lequel je butais. Ainsi le théorème 1 est-il prouvé en général [2].

On peut même donner au théorème 1 une forme un peu plus générale.

THÉORÈME 2. *Soient $\Gamma_1,\ldots,\Gamma_k$ des arcs de Jordan réguliers, disjoints, allant à l'infini dans $\mathbb{C}^n$ et ε une fonction continue strictement positive sur la réunion de ces arcs, alors quelle que soit la fonction f continue sur la réunion de ces arcs il existe g entière sur $\mathbb{C}^n$ telle* *que* $|f(z)-g(z)|<\varepsilon(z)$, *pour tout* z *appartenant à l'un des* Γ_i .

De cela ou peut déduire un résultat sur le comportement des fonctions holomorphes à la frontière, à savoir:

COROLLAIRE. *Si* Γ_0 *et* Γ_∞ *sont deux arcs de Jordan réguliers, disjoints, allant à l'infini dans* $\mathbb{C}^n$, *il existe une fonction g entière sur* $\mathbb{C}^n$ *telle que* $g(z) \to 0$ *sur* Γ_0 *à l'infini et* $g(z) \to \infty$ *sur* Γ_∞ *à l'infini.*

S. Scheinberg [9], en suivant des méthodes d'analyse classique, a pu montrer que toute fonction continue sur $\mathbb{R}^n$ est limite uniforme sur $\mathbb{R}^n$ de fonctions entières sur $\mathbb{C}^n$. En utilisant le théorème 1 on peut obtenir la généralisation suivante:

THÉORÈME 3. *Soient* $\Gamma_1, \Gamma_2, \ldots, \Gamma_k$ *des arcs de Jordan réguliers compacts ou allant à l'infini respectivement dans* $\mathbb{C}^{n_1}, \ldots, \mathbb{C}^{n_k}$. *Posons* $V = \Gamma_1 \times \ldots \times \Gamma_k$ *qui est contenu dans* $\mathbb{C}^n$, *avec* $n = n_1 + \ldots + n_k$, *et soit* ε *une fonction continue strictement positive sur V. Si f est continue sur V il existe g entière sur* $\mathbb{C}^n$ *telle que* $|f(z) - g(z)| < \varepsilon(z)$, *pour tout z de V.*

Probablement les même résultats, où les arcs sont supposés localement rectifiables, sont aussi vrai à condition de remplacer la proposition de Stolzenberg par l'énoncé analogue avec l'hypothèse de rectifiabilité. Mais la démonstration, basée sur les idées de H. Alexander [1], en est certainement très technique.

CAS DU THÉORÈME 1. La démonstration de T. Carleman utilise la représentation intégrale du Cauchy pour certains domaines de $\mathbb{C}$, mais pour $n \geq 2$ le même argument utilisant l'une des représentation intégrales de Bochener-Martinelli ou Henkin ne semble pas fonctionner, ainsi a-t-il fallu utiliser un argument du type Merglyan pour des domaines particuliers de $\mathbb{C}^n$. Le premier lemme géométrique est bien connu, la proposition suivante -très difficile- est due à G. Stolzenberg [10] (voir aussi [11], chapitre 6, §30).

LEMME 1. *Soient E un espace de Banach réel pour la norme* $\| \ \|$,

F un sous-espace vectoriel dense dans E et $\phi_1,\ldots,\phi_n$ *des formes linéaires continue sur E. Quels que soient* $\varepsilon>0$ *et x dans E il existe y dans F tel que* $||z-y||<\varepsilon$ *et* $\phi_i(x) = \phi_i(y)$, *pour* $i=1,\ldots,n$.

Démonstration. On fait un raisonnement par récurrence. Si n=1, d'après la densité il existe a,b dans F tels que $||x-a||<\varepsilon$, $||x-b||<\varepsilon$, $\phi_1(a)>\phi_1(x)$, $\phi_1(b)<\phi_1(x)$. On prend $y=ta+(1-t)b$, avec $0\leq t\leq 1$ convenable.□

Si X est un compact de $\mathbb{C}^n$, $\hat{X}$ dénote l'enveloppe polynomialement convexe de X. C(X) dénote l'algèbre des fonctions continues sur X, P(X) dénote l'algèbre des fonctions continue sur X uniformément approximables par des polynômes et R(X) dénote l'algèbre des fonctions continues sur X uniformément approximables par des fonctions rationnelles holomorphes sur X. En plus $\check{H}^1(X,\mathbb{Z})$ est le premier groupe de cohomologie de Čech à coefficients entiers.

THÉORÈME (Stolzenberg). *Soient X un compact polynomialement convexe de* $\mathbb{C}^n$ *et K une réunion finie d'arcs réguliers de* $\mathbb{C}^n$, *alors on a les propriétés suivantes:*

- 1° $(K\cup X)\hat{}\setminus(K\cup X)$ *a une structure de sous-ensemble analytique (éventuellement vide) de dimension 1 de* $\mathbb{C}^n\setminus(K\cup X)$.
- 2° $K\cup X$ *est rationnellement convexe.*
- 3° *Si* $g\in C(K\cup X)$ *et si* $g_{|X}\in P(X)$ *alors* $g\in R(K\cup X)$.
- 4° *Si l'application de* $\check{H}^1(K\cup X,\mathbb{Z})$ *dans* $\check{H}^1(X,\mathbb{Z})$ *définie par l'inclusion de X dans* $K\cup X$ *est injective alors* $K\cup X$ *est polynomialement convexe.*

Ce théorème implique évidemment (avec $X=\emptyset$) l'approximation polynômiale sur les arcs de Jordan compacts réguliers. D'après le théorème de Arens-Royden, la condition 4° est équivalente à dire que: si une fonction continue sur $K\cup X$ n'est pas nulle sur cet ensemble et a un logarithme continu sur X alors elle a un logarithme continu sur $K\cup X$.

LEMME 2. *Soient X un compact polynomialement convexe et* α,β *deux* *arcs disjoints réguliers de* $\mathbb{C}^n$ *tels que* $\alpha\cap X$ *et* $\beta\cap X$ *aient respectivement un seul point, alors:*

-1° $X\cup\alpha\cup\beta$ *est polynomialement convexe.*

-2° *Si* $g\in C(X\cup\alpha\cup\beta)$ *et* $g\equiv 0$ *sur* X *alors* $g\in P(X\cup\alpha\cup\beta)$, *autrement dit pour* $\varepsilon>0$ *il existe un polynôme* p *tel que* $|g(z)-p(z)|<\varepsilon$ *sur* $X\cup\alpha\cup\beta$ *et tel que* $g(z)=p(z)=0$ *pour* $z\in(\alpha\cup\beta)\cap X$.

Démonstration. D'après les hypothèses, toute fonction continue sur $X\cup\alpha\cup\beta$ qui ne s'annule pas sur cet ensemble et qui admet un logarithme continu sur X admet aussi un logarithme continu sur $X\cup\alpha\cup\beta$. Ainsi d'après le 4° du théorème de Stolzenberg, $X\cup\alpha\cup\beta$ est polynomialement convexe. D'après le 3°, $g\in R(X\cup\alpha\cup\beta)$, donc, d'après ce qui précède et le théorème d'Oka-Weil, $g\in P(X\cup\alpha\cup\beta)$. En appliquant le lemme 1 aux formes linéaires qui sont les évaluations aux points d'intersection respectifs de α et β avec X on obtient la condition finale.□

Nous dénoterons par Γ un arc régulier allant à l'infini dans les deux directions et nous supposerons que Γ contient l'origine. Pour $r>0$ nous dénoterons par $B(r)$ la boule ouverte du centre 0 de rayon r, par $\gamma(r)$ le sous-arc de $\Gamma\cap B(r)$ contenant 0 -auquel cas $\gamma(r)$ a deux extrêmités sur $\partial B(r)$- et par $\sigma(r)$ le sous-arc complémentaire dans Γ des deux composantes connexes non bornées de $\Gamma\setminus B(r)$. Il est clair que $\gamma(r)$ et $\sigma(r)$ sont deux arcs ouverts bornés de $\mathbb{C}^n$.

On définit une suite (r_k) par récurrence, de façon que $r_1=1$, $r_k > r_{k-1}+1$ et que:

(a) $\sigma(r_{k-1})\subset B(r_k)$

(b) $(\bar{B}(r_{k-1})\cup\bar{\sigma}(r_{k-1}))\hat{}$ et $\bar{\sigma}(r_k)\setminus\gamma(r_k)$ soient disjoints.

(a) est évident puisque $\sigma(r)$ est borné, (b) résulte du fait que le premier ensemble $(\bar{B}(r_{k-1})\cup\bar{\sigma}(r_{k-1}))\hat{}$ est compact et que l'arc s'éloigne à l'infini, donc en choisissant r_k assez grand on peut toujours s'arranger pour que le deuxième ensemble soit disjoint du premier. La suite (r_k) étant construite on notera γ_k pour $\gamma(r_k)$, σ_k pour $\sigma(r_k)$ et B_k pour $B(r_k)$.

Pour $k\geq 2$ posons $X_k = (\bar{B}_{k-2}\cup\bar{\gamma}_{k-1})^\wedge$ et $Y_k = X_k\cup\bar{\gamma}_k$. Comme X_k est une enveloppe polynomialement convexe, il est polynomialement convexe.

L'importance du lemme qui, suit dû à H. Alexander [2], est de donner une description plus explicite de ces deux ensembles.

LEMME 3.

-1° $X_k = (\bar{B}_{k-2}\cup\bar{\sigma}_{k-2})^\wedge\cup(\gamma_{k-1}\setminus\sigma_{k-2})$

-2° $Y_k = X_k\cup\alpha_k\cup\beta_k$, *où* α_k,β_k *sont des arcs réguliers disjoints qui rencontrent* X_k *respectivement en un seul point.*

Démonstration: 1° Dénotons par T_k le second membre de l'inclusion du 1°. Comme $X_k = ((\bar{B}_{k-2}\cup\bar{\sigma}_{k-2})\cup(\bar{\gamma}_{k-1}\setminus\sigma_{k-2}))^\wedge$ on obtient que $X_k = \hat{X}_k\supset(\bar{B}_{k-2}\cup\bar{\sigma}_{k-2})^\wedge\cup(\bar{\gamma}_{k-1}\setminus\sigma_{k-2})^\wedge$, mais $\bar{\gamma}_{k-1}\setminus\sigma_{k-2}$est un arc régulier, donc polynomialement convexe, ainsi $T_k\subset X_k$ et l'autre inclusion $X_k\subset\hat{T}_k$ est évidente. Il suffit de montrer que T_k est polynomialement convexe. Raisonnons par l'absurde. D'après le 1° du théorème de Stolzenberg, $\hat{T}_k\setminus T_k$ est une sous-variété analytique de dimension 1 de $\mathbb{C}^n\setminus T_k$. Soit V une composante analytique non vide et irréductible de $\hat{T}_k\setminus T_k$. Commençons par montrer que $\bar{V}\setminus(\bar{B}_{k-2}\cup\bar{\gamma}_{k-1})$ est une sous-variété analytique de $\mathbb{C}^n\setminus(\bar{B}_{k-2}\cup\bar{\gamma}_{k-1})$. D'après la définition de T_k, il suffit de vérifier localement en $x\in\bar{V}\cap Q$, où $Q=(\bar{B}_{k-2}\cup\bar{\sigma}_{k-2})^\wedge\setminus(\bar{B}_{k-2}\cup\bar{\sigma}_{k-2})$; en effet si $x\in V\setminus(\bar{B}_{k-2}\cup\bar{\gamma}_{k-1})$ il n'y a pas de problème et si $x\in\bar{V}\setminus(\bar{B}_{k-2}\cup\bar{\gamma}_{k-1})$ avec $x\in T_k$ alors $x\in\bar{\gamma}_{k-1}\setminus\sigma_{k-2}$ est impossible donc $x\in(\bar{B}_{k-2}\cup\bar{\sigma}_{k-2})^\wedge$. D'après le 1° du théorème de Stolzenberg, X_k et Q ont une structure analytique au voisinage de x, de plus on voit facilement que $\bar{V}\subset X_k$, $V\subset X_k\setminus Q$ et $Q\subset X_k$. Il résulte que près de x, $\bar{V}$ est l'union de quelques composantes analytiques locales de X_k en x, en fait près de x, $\bar{V} = V\cup\{x\}$. Posons $W = \bar{V}\setminus(\bar{B}_{k-2}\cup\bar{\gamma}_{k-1})$, c'est une sous-variété irréductible de $\mathbb{C}^n\setminus(\bar{B}_{k-2}\cup\bar{\gamma}_{k-1})$ et de plus $\bar{W}\setminus W\subset\bar{B}_{k-2}\cup\bar{\gamma}_{k-1}$. D'après le principe du maximum $\bar{W}\subset(\bar{B}_{k-2}\cup\bar{\gamma}_{k-1})^\wedge$. Soit $p\in V\subset W$, comme $p\notin T_k$, $p\notin(\bar{B}_{k-2}\cup\bar{\sigma}_{k-2})^\wedge$ et donc il existe un polynôme h tel que $h(p)=0$ et $\mathrm{Re}\, h(z) < 0$ sur $(\bar{B}_{k-2}\cup\bar{\sigma}_{k-2})^\wedge$. Alors h(W) est un voisinage de 0 ou sinon $h\equiv 0$ sur W. Dans le dernier cas $h\equiv 0$ sur $\bar{W}$ et donc $\bar{W}\setminus W$ est disjoint de $\bar{B}_{k-2}\cup\bar{\sigma}_{k-2}$, ce qui avec $\bar{W}\setminus W\subset(\bar{B}_{k-2}\cup\bar{\sigma}_{k-2})\cup(\bar{\gamma}_{k-1}\setminus\sigma_{k-2})$ donne $\bar{W}\setminus W\subset\bar{\gamma}_{k-1}\setminus\sigma_{k-2}$ donc $\bar{W}\subset(\bar{\gamma}_{k-1}\setminus\sigma_{k-2})\hat{}$; d'après le principe du maximum, mais $\bar{\gamma}_{k-1}\setminus\sigma_{k-2}$ est polynomialement convexe car réunion disjointe de deux arcs de Jordan réguliers. On peut donc supposer que h(W) et un voisinage de 0, comme $h(\bar{\gamma}_{k-1}\setminus\sigma_{k-2})$ est non dense dans $\mathbb{C}$ il existe $\alpha\in h(W)$ petit tel que $h\neq\alpha$ sur $\bar{\gamma}_{k-1}\setminus\sigma_{k-2}$. Posons $g=h-\alpha$, si α est choisi assez petit alors $\mathrm{Re}\,g<0$ sur $(\bar{B}_{k-2}\cup\bar{\sigma}_{k-2})\hat{}$; $g(q)=0$ pour un $q\in W$, $g\neq 0$ sur $\bar{\gamma}_{k-1}\setminus\sigma_{k-2}$. En appliquant la première relation, le polynôme g a un logarithme continu sur $(\bar{B}_{k-2}\cup\bar{\sigma}_{k-2})\hat{}$ donc sur $\bar{B}_{k-2}\cup\bar{\sigma}_{k-2}$. D'après la troisième relation on peut prolonger ce logarithme sur les deux arcs formés par $\bar{\gamma}_{k-1}\setminus\sigma_{k-2}$, donc g a un logarithme continu sur $\bar{B}_{k-2}\cup\bar{\sigma}_{k-2}\cup(\bar{\gamma}_{k-1}\setminus\sigma_{k-2})$ qui contient $\bar{W}\setminus W$. En appliquant le principe d'argument à g sur W on obtient une contradiction avec la deuxième relation $g(q)=0$.

2° On écrit $\bar{\gamma}_k\setminus\gamma_{k-1}$ comme l'union disjointe de deux arcs α_k et β_k joignant ∂B_{k-1} à ∂B_k. En fait $X_k\cap\alpha_k$ est l'ensemble $\partial\alpha_k\cap\partial B_{k-1}$ réduit à un seul point, de même pour β_k, pour la raison suivante. On a $\alpha_k\cap B_{k-1}\subset(\bar{\gamma}_k\setminus\gamma_{k-1})\cap B_{k-1}\subset\bar{\sigma}_{k-1}\setminus\gamma_{k-1}$ et d'après le 1° on a $X_k=(\bar{B}_{k-2}\cup\bar{\sigma}_{k-2})\hat{}\cup\bar{\gamma}_{k-1}$. En utilisant le propriété b) on obtient que $X_k\cap\alpha_k\cap B_{k-1}$ est vide. Alors $X_k\cap\alpha_k\subset\partial B_{k-1}$ donc $X_k\cap\alpha_k=\partial\gamma_{k-1}\cap\alpha_k=\partial\alpha_k\cap\partial B_{k-1}$, ce qu'il fallait montrer. □

Nous pouvons maintenant démontrer le théorème 1.

Posons $\eta_n=\mathrm{Min}\ \varepsilon(z)$, pour $z\in\Gamma\cap\bar{B}_n$, on peut construire facilement une suite (ε_n) telle que

$$\sum_{k=1}^{n}\varepsilon_k\le\eta_n \qquad (1)$$

Comme $C(\bar{\gamma}_1)=P(\bar{\gamma}_1)$ il existe un polynôme p_1 tel que

$$\sup_{\gamma_1}|f-p_1|<\varepsilon_1 \qquad (2)$$

$$f=p_1 \quad \text{sur} \quad \partial\gamma_1$$

On définit g_2 sur Y_2 par $g_2=0$ sur X_2 et $g_2=f-p_1$ sur $\bar{\gamma}_2\setminus\gamma_1=\alpha_2\cup\beta_2$. D'après (2), g_2 est continue sur Y_2. En appliquant le lemme 2 à g_2 sur Y_2 on obtient un polynôme p_2 tel que $\sup_{Y_2}|g_2-p_2|<\varepsilon_2$ et $g_2=p_2$ sur $(\partial\alpha_2\cup\partial\beta_2)\cup\partial B_2$. On procède maintenant par récurrence en supposant connus

$p_1,\ldots,p_{n-1}$ et les fonctions $g_2,\ldots,g_{n-1}$ avec $g_k \in C(Y_k)$ satisfaisant à:

$$\left\{\begin{array}{ll} g_k = 0 & \text{sur } X_k \\ g_k = f - \sum_{i=1}^{k-1} p_i & \text{sur } \alpha_k \cup \beta_k \\ |g_k - p_k| < \varepsilon_k & \text{sur } Y_k \\ g_k = p_k & \text{sur } (\partial\alpha_k \cup \partial\beta_k) \cap \partial\beta_k \end{array}\right. \qquad (3)_k$$

pour $2 \le k \le n-1$. On peut alors définir g_n par $g_n = 0$ sur X_n et $g_n = f - \sum_{i=1}^{n-1} p_i$ sur $\alpha_n \cup \beta_n$, qui est continue. On peut alors, d'après le lemme 2, construire un polynôme p_n vérifiant les deux dernières relations de $(3)_n$. En posant $g = \sum_{n=1}^{\infty} p_n$, il est standard de vérifier que g est entière sur $\mathbb{C}^n$ et vérifie $|f(z)-g(z)| < \varepsilon(z)$ sur Γ.□

CAS DU THÉORÈME 2. Sa démonstration se fait de façon semblable à celle du théorème 1.

Pour la démonstration du corollaire on prend $f(z)=0$ si z est sur Γ_0 et $f(z) = ||z||$ si z est sur Γ_∞, ainsi définie g est continue sur $\Gamma_0 \cup \Gamma_\infty$. Posons $\varepsilon(z) = 1/||z||$, en supposant par exemple que 0 n'est pas dans $\Gamma_0 \cup \Gamma_\infty$, alors, d'après le théorème 2 il existe g entière telle que

$$|g(z)| < 1/||z|| \qquad \text{sur } \Gamma_0$$

$$|g(z) - ||z||| < 1/||z|| \qquad \text{sur } \Gamma_\infty$$

d'où le résultat. □

CAS DU THÉORÈME 3. Donnons sa démonstration. Supposons pour l'instant que $\varepsilon(z)$ est constante et égale à $\varepsilon > 0$. On commence par écrire $f = \sum_{i=1}^{\infty} f_i$, sur V, où les f_i sont à supports compacts vérifiant la propriété suivante: si K est un compact de $\mathbb{C}^n$ on peut supposer que

Supp $f_i \cap K = \emptyset$ pour i assez grand. Si on arrive à démontrer le théorème pour une fonction continue sur V à support compact le théorème sera démontré, car pour chaque f_i on choisit g_i entière sur $\mathbb{C}^n$ telle que $|f_i(z) - g_i(z)| < \varepsilon/2^i$ sur V, il suffit alors de remarquer que $g = \sum_{i=1}^{\infty} g_i$ converge uniformément sur tout compact de $\mathbb{C}^n$ et vérifie $|f(z)-g(z)| < \varepsilon$

sur V. Supposons donc maintenant f continue sur V et à support compact, alors $F \in C_0(V)$, l'algèbre des fonctions continues sur V qui s'annulent à l'infini. Si on considère la sous-algèbre des fonctions de la forme $h = \sum_{j=1}^{N} \phi_1^j(z_1)\ldots\phi_k^j(z_k)$, restreintes à V, où les ϕ_i^j sont continues et à support compact dans Γ_i, on constate que cette sous-algèbre est auto-conjuguée, sépare les points y compris le points à l'infini de l'espace localement compact V donc, d'après le théorème de Stone-Weierstrass, il existe h ayant cette propriété telle que $|f(z)-h(z)|< \varepsilon/2$ sur V. Maintenant il suffit de montrer que h est approximable $\varepsilon/2$ près, par une fonction entière, mais pour chaque (i,j) il existe g_i^i entière en $z_i \in \mathbb{C}^{n_i}$ telle que:

$$|\phi_i^j(z_i)-g_i^j(z_i)|<\varepsilon_i^j \tag{4}$$

où les $\varepsilon_i^j > 0$ vérifient

$$\prod_{i=1}^{k} (\varepsilon_i^j + \|\phi_i^j\|) - \prod_{i=1}^{k} \|\phi_i^j\| < \varepsilon/2N \tag{5}$$

Donc en posant $g=\sum_{j=1}^{N} g_1^j(z_1)\ldots g_k^j(z_k)$ on obtient $|f(z)-g(z)|<\varepsilon$ pour $z \in V$.

Si maintenant $\varepsilon(z)$ est une fonction de $z \in V$, d'après ce qui précède il existe h entière sur $\mathbb{C}^n$ telle que $|\mathrm{Log}\,\varepsilon(z)-h(z)|<1$ sur V. Posons $h_1(z)= e^{h(z)-1}$ qui est entière et non nulle. Alors $|h_1(z)|<\varepsilon(z)$ pour $z \in V$, il existe g_1 entière telle que $|\frac{f(z)}{h_1(z)} - g_1(z)|<1$ donc $|f(z)-g(z)|< |h_1(z)|<\varepsilon(z)$, avec $g(z)=h_1(z)g_1(z)$.□

BIBLIOGRAPHIE

[1] H. Alexander, Polynomial approximation and hulls of sets of finite linear measure in $\mathbb{C}^n$, *Amer. J. Math.* 93 (1971), 65-75.

[2] H. Alexander, A Carleman theorem for curves in $\mathbb{C}^n$,*Math. Scand.*, à paraître.

[3] N.U. Arakelian, Approximation uniforme et asymptotique par des fonctions entières sur les ensembles fermés non bornés (en russe).

Dokl, Akad. Nauk SSSR 157 (1964), 9-11.

[4] T. Carleman, Sur un théorème de Weirstrass, *Ark. för Math. Astr. Fys.* 20 (1927), 1-5.

[5] W.H.J. Fuchs, *Théorie de l'approximation des fonctions d'une variable complexe*, Presses de l'Université de Montréal, Montréal, 1968.

[6] M. Keldych et M. Lavrentieff, Sur un problème de M. Carleman, *Dokl. Akad. Nauk SSSR*. 23 (1938), 746-748.

[7] J. Nunemacher, Approximation theory on totally real submanifolds, *Math. Ann.* 224 (1976), 129-141.

[8] J. Nunemacher, Approximation theory on CR-submanifolds, *Proc. Symp. Pure Math.* 30 (1977), 181-186.

[9] S. Scheinberg, Uniform approximation by entire functions, *J. d'Analyse Math.* 29 (1976), 16-18.

[10] G. Stolzenberg, Uniform approximation on smooth curves, *Acta Math.* 115 (1966), 185-198.

[11] E.L. Stout, *The theory of uniform algebras*. Bogden and Quigley, Tarrytown-on-Hudson, New York, 1971.

[12] E.L. Stout, Uniform approximation on certain unbounded sets, dans ce livre.

[13] J. Wermer, *Banach algebras and Several Complex variables*. 2nd edition. Springer-Verlag, New York, 1976.

Département de Mathématiques
Université Laval
Québec.

AN ALGEBRAIC QUESTION RELATED TO THE FUNCTION THEORY FOR REAL SUB-MANIFOLDS OF $\mathbb{C}^n$

by

Gary A. Harris*

INTRODUCTION. The purpose of the following discussion is to advertise a recent result of the author which has application the the problems of local holomorphic approximation and local holomorphic extension for C^∞ real submanifolds of $\mathbb{C}^n$ with CR singularity and to pose a geometric question motivated by this result. Section 1 contains the above mentioned result and applications along with a minimum amount of terminology. The related geometric question is contained in section 2, as well as, a pertinent example. In addition, section 2 contains an algebraic reformulation of this geometric question in the real- analytic category, with all necessary algebraic concepts and terms defined.

This paper is a somewhat expanded version of the talk given by the author the Conférence sur l'approximation complexe. The author is grateful to the conference organizers for providing such a "forum".

1. THE RESULT. As indicated above, we will be interested only in local questions. Thus assume M is a C^∞ real submanifold of $\mathbb{C}^n$ with global defining functions in a neighborhood of the origin, 0. That is, $M = \{Z \varepsilon U | \rho_1(Z) = \ldots = \rho_m(Z) = 0\}$ where

(i) U is an open subset of $\mathbb{C}^n$ containing 0.

(ii) For all $1 \le i \le m$, $\rho_i : U \to \mathbb{R}$ is C^∞

(iii) For all $1 \le i \le m$, $\rho_i(0) = 0$.

---

* Partially supported by the National Science Foundation

(iv) For any $p \varepsilon M, d_p\rho_1 \wedge \ldots \wedge d_p\rho_m \neq 0$.

Condition (iii) is a technical convenience which is used in section 2 and condition (iv) is standard: the Fréchet differentials are linearly independent at any point of M.

DEFINITION.

(a) M is CR provided $rk[\partial\rho_i/\partial\bar{Z}_j(p)]$ is independent of $p \varepsilon M$.

(b) If M is CR and $rk[\partial\rho_i/\partial\bar{Z}_j(\cdot)] = s$ then M is *generic in* $\mathbb{C}^n$ provided $s = \min\{m,n\}$.

(c) $E(M) = \{p \varepsilon M \mid rk[\partial\rho_i/\partial\bar{Z}_j(.)]$ is not constant in any neighborhood of $p\}$ is called the *CR singular set* of M.

(d) $CR(M) = M \setminus E(M)$

(e) M, not nessarily CR, is *generic in* $\mathbb{C}^n$ provided CR(M) is CR and generic in $\mathbb{C}^n$.

For every positive integer k there is a k-th order tangential partial differential operator, $\bar{\partial}_k$, defined on M with $\bar{\partial}_1$ being the usual tangential Cauchy-Riemann operator. (See [4] for a detailed description of $\bar{\partial}_k$.) in [4] the following theorem is proved.

THEOREM A. *Suppose $E(M)$ is nowhere dense in M, $m \leq n$, and M is generic in $\mathbb{C}^n$. Moreover suppose $f: U \to \mathbb{C}$ is C^∞ with $\bar{\partial} f \varepsilon \mathcal{O}^{k-1}(M)$. (i.e. The complex-conjugate linear part of the Frechet differential of f vanishes identically on M to order k-1.) There exists a C^∞ function $g: U \to \mathbb{C}$ such that $g - f \varepsilon \mathcal{O}(M)$ and $\bar{\partial} g \varepsilon \mathcal{O}^k(M)$ if and only if $\bar{\partial}_k f = 0$.*

The notation used here is not precise and the reader should consult [4] for a technically precise statement of this result. However the "flavor" of the theorem is evident. If k=1 the hypothesis $\bar{\partial} f \varepsilon \mathcal{O}^0(M)$ is vacuous, thus one has a condition to guarantee the existence of a extension whose $\bar{\partial}$ vanishes to order 1. Then one has a condition that this new function has an extension whose $\bar{\partial}$ vanishes to order 2, etc. Thus a well known argument using some L^2-estimates of L. Hörmander (See Hörmander-Wermer [5], R. Nirenberg-Wells [6], or Freeman [1].) implies the following approximation result. [4]

THEOREM B. *If U is a domain of holomorphy, K a compact subset of M (M as in Theorem A) which is holomorphically convex with respect to U, and $f:U\to\mathbb{C}$ a C^∞ function such that $\bar{\partial}_k f = 0$ for all positive integers k, then f is the uniform limit on K of holomorphic functions on U.* Again one should consult [4] for a technically precise statement of the conditon "$\bar{\partial}_k f = 0$".

If M is a real-analytic submanifold then the condition "$E(M)$ is nowhere dense" is redundant and one has the following theorem. [4]

THEOREM C. *If M is real-analytic, generic in $\mathbb{C}^n$, $m\le n$, and f a real-analytic function near 0, then there is a holomorphic function F near 0 such that $F-f \in \mathcal{O}(M)$ if and only if $\bar{\partial}_k f = 0$ for all positive integers k.*

Theorem C does not necessarily yield a holomorphic function defined on all of U and thus is more local in character than either Theorem A of Theorem B. This local feature in the real-analytic category is exploited to get the algebraic formulation of the question which begins section 2.

2. THE QUESTION. The condition "$m\le n$ and M generic in $\mathbb{C}^n$" is necessary for the arguments used in [4] to prove theorems A, B, and C. Thus a "natural" geometric question arises.

QUESTION A. Given M do there exist an integer k, neighborhoods U and V of 0 in $\mathbb{C}^n$, and a biholomorphic mapping $H:U\to V$ such that

(1) $k\le\min\{2n-m,n\}$

(2) $H(M)\subset\mathbb{C}^k\times\{0\}$

(3) $H(M)$ is generic in $\mathbb{C}^k$?

Because of (2) one may think of $H(M)$ as being contained in $\mathbb{C}^k$ and (1) is the condition that the real codimension of $H(M)$ in $\mathbb{C}^k$ is less than or equal to k.

DEFINITION. If such k and H exist then $\tilde{M} \equiv H^{-1}((\mathbb{C}^k\times\{0\})\cap V)$ is called the *minimal complex envelope of M* near 0.

If M is real-analytic and CR then it follows from results of Tomassini [7] that the minimal complex envelope always exists locally. However this is not necessarily the case if $E(M) \neq \emptyset$.

EXAMPLE. Suppose $M \subset \mathbb{C}^3$ is defined by

$$\rho_1 = \tfrac{1}{2}(Z_2+\bar{Z}_2) - \frac{1}{4i}(Z_1^2-\bar{Z}_1^2)$$

$$\rho_2 = \frac{1}{2i}(Z_2-\bar{Z}_2) + \tfrac{1}{4}(Z_1-\bar{Z}_1)^2$$

$$\rho_3 = \tfrac{1}{2}(Z_3+\bar{Z}_3) - \frac{1}{4i}(Z_1^2-\bar{Z}_1^2)\,e^{\frac{1}{2i}(Z_1-\bar{Z}_1)}$$

$$\rho_4 = \frac{1}{2i}(Z_3-\bar{Z}_3) + \tfrac{1}{4}(Z_1-\bar{Z}_1)^2\,e^{\frac{1}{2i}(Z_1-\bar{Z}_1)}$$

For reasons which will follow, M is a uniqueness set for holomorphic functions of $\mathbb{C}^3$ near 0. (i.e. If f is holomorphic near 0 and $f \in \mathcal{O}(M)$ then $f=0$. Thus there is no biholomorphic H near 0 on $\mathbb{C}^3$ such that $H(M) \subset \mathbb{C}^2 \times \{0\}$ even though $2n-m = 2$ in this case. Moreover for this example it is easy to see that $E(M)=\{0\}$, the CR singularity is isolated.

As indicated by the above example, the question is interesting even in the real-analytic category. Thus the submanifold M and all functions will henceforth be assumed to be real-analytic in some neighborhood of 0. Suppose W is a neighborhood of 0 in $\mathbb{R}^{2n-m}$, $\Phi \equiv (\phi_1,\ldots,\phi_n)$ $: W \to \mathbb{C}^n$ is a real-analytic parametrization of M, and $\Phi(0) = 0$. Let $\tilde{\Phi} \equiv (\tilde{\phi}_1,\ldots,\tilde{\phi}_n)$ be the unique holomorphic mapping which makes the following diagram commute:

$$\begin{array}{ccc} W \subset \mathbb{R}^{2n-m} & \xrightarrow{\ \Phi\ } & \mathbb{C}^n \\ \cap \downarrow & \circlearrowleft & \nearrow \tilde{\Phi} \\ \mathbb{C}^{2n-m} & & \end{array}$$

This approach is developed in [3] where the following facts are derived.

(1) M is CR near 0 if and only if $\operatorname{rk}[\partial\tilde{\phi}_i/\partial Z_j(.)]$ is constant near 0.

(2) M (not necessarily CR) is generic in $\mathbb{C}^n$ and $m \leq n$ if and only if $\operatorname{rk}[\partial\tilde{\phi}_i/\partial Z_j] = n$

(3) For f holomorphic on $\mathbb{C}^n$ near 0, $f \in \mathcal{O}(M)$ if and only if $f \circ \tilde{\Phi} \equiv 0$.

The expression $rk[\partial\tilde{\phi}_i/\partial Z_j]$ in (2) means the generic rank of the matrix of functions (i.e. the maximum rank of $[\partial\tilde{\phi}_i/\partial Z_j(.)]$ near 0).

Let $\mathbb{C}_n$ denote the ring of convergent power series centered at 0 with complex coefficients and n complex indeterminates (i.e. the germ of holomorphic functions at 0). Let $\tilde{\Phi}_*:\mathbb{C}_n\to\mathbb{C}_k$ be the ring homomorphism given by $\tilde{\Phi}_*(f)\equiv f\circ\tilde{\Phi}$. Thus (3) becomes

(4) $f\in\mathcal{O}(M)$ if and only if $f\in\ker\tilde{\Phi}_*$.

In the example, "M is a uniqueness set for holomorphic functions near 0" follows because if $\tilde{\Phi}\equiv(Z,ZW,ZWe^W)$ then $\tilde{\Phi}_*:\mathbb{C}_3\to\mathbb{C}_2$ is injective. (See [3, Section 6] and Grauert-Remmert[2, pg 121].)

The previous geometric question can now be stated algebraically. The "$\sim$" notation will be omitted as all maps are now assumed to be holomorphic.

QUESTION B. Suppose $\Phi=(\phi_1,\ldots,\phi_n)$ and for all $1\le i\le n$, $\phi_i\in\mathbb{C}_k$ and $\phi_i(0)=0$. Further suppose $rk[\partial\phi_i/\partial Z_j]=s$. Does there exist $H\equiv(H_1,\ldots,H_n)$ with each $H_i\in\mathbb{C}_n$ and $d_0H_1\wedge\ldots\wedge d_0H_n\ne0$ such $\{H_{s+1},\ldots,H_n\}\subset\ker\Phi_*$?

If such H exists then $M\subset V(H_{s+1},\ldots,H_n)$ where $V(H_{s+1},\ldots,H_n)\equiv\{Z \text{ near } 0\,|\,H_{s+1}(z)=\ldots=H_n(Z)=0\}$ is the germ of a regular complex subvariety of complex dimension s and is , after a linear coordinate change, the graph near 0 of a holomorphic mapping $G:\mathbb{C}^s\to\mathbb{C}^{n-s}$. Thus the question can be restated.

QUESTION C. Suppose Φ is as above. Do there exist coordinates for $\mathbb{C}^n$ near 0 and $G_{s+1},\ldots,G_n\in\mathbb{C}_s$ such that for all $1\le j\le n-s$, $G_{s+j}(0)=0$ and $\phi_{s+j}=G(\phi_1,\ldots,\phi_s)$?

(Here ϕ_i is the ith coordinate of Φ in the new coordinate system.)

If M is CR, i.e. $rk[\partial\phi_i/\partial Z_j(.)]$ is constant, then the answer to Question C is "yes" by the holomorphic implicit function theorem. Thus the existence of the local minimal complex envolope follows. The question is asking for a generalized implicit function theorem is the case of non-constant rank. This can be made more precise as follows.

DEFINITION. (Grauert-Remmert [2]) Suppose $\mathcal{O}$ is an ideal $\mathbb{C}_n$ and M is the maximal ideal in $\mathbb{C}_n$. Let $\delta: M \to M/M^2$ be the natural mapping. The Jacobian Rank of $\mathcal{O}$, denoted jg $\mathcal{O}$, is the dimension over $\mathbb{C}$ of the finite dimensional vector space $\delta(\mathcal{O})$.

An easy exercise proves the following theorem.

THEOREM D. Suppose Φ is the holomorphic mapping associated with M as above.

$$n-S=jg(\ker \Phi_*) \Leftrightarrow \begin{cases} M \subsetneq \mathbb{C}^S \\ M \not\subset \mathbb{C}^{S-1} \end{cases}$$

Because of Theorem D the above question has the following algebraic formulation.

QUESTION D. Given a ring homorphism $\Phi_*: \mathbb{C}_n \to \mathbb{C}_k$, under what conditions does

$$jg(\ker \Phi_*) = n - rk\ \Phi_* \ ?$$

(Here rk $\Phi_* \equiv$ rk$[\partial\phi_i/\partial Z_j]$ where for all $1 \le i \le n$ $\phi_i \equiv \Phi_*(W_i)$ and W_i are coordinates for $\mathbb{C}^n$.

In general $jg(\ker \Phi_*) \le n - rk\ \Phi_*$, with the above example yielding strict inequality. A related problem of obvious geometric interpretation is that of determining $jg(\ker \Phi_*)$ in general from the coordinate functions, ϕ_i.

REFERENCES

[1] M. Freeman, Tangential Cauchy-Riemann Equations and Uniform Approximation, *Pacific J. Math.* 33 (1970) 101-108.

[2] H. Grauert and R. Remmert, *Analytische Stellenalgebren*, Springer Verlag, New York, 1970.

[3] G. Harris, The Traces of Holomorphic Functions on Real Submanifolds of $\mathbb{C}^n$, *Trans. Amer. Math. Soc.* (to appear).

[4] G. Harris, Higer Order Analogues to the Tangential Cauchy-Riemann Equations for Real Submanifolds of $\mathbb{C}^n$ with CR Singularity, *Proc. Amer. Math. Soc.* (to appear).

[5] L. Hörmander and J. Wermer, Uniform Approximation on Compact Subsets of $\mathbb{C}^n$, *Math. Scan.* 23 (1968), 5-21.

[6] R. Nirenberg and R. O. Wells, Jr., Approximation Theorems on Differentiable Submanifolds of a Complex Manifold, *Trans. Amer. Math. Soc.* 142 (1969), 15-35.

[7] G. Tomassini, Trace delle Funzioni Olomorfe sulla Sottovarieta Analiticha Reali d'una Varieta Complessa, *Ann. Scuola Norm. Sup. Pisa* (1966), 31-43.

Department of Mathematics
Texas Tech. University

APPROXIMATION AND NON-APPROXIMATION ON RIEMANN SURFACES

by

Stephen Scheinberg*

My purpose in both the talk and this article is to acquaint the audience and readers with just a few of the developments, both historical and recent, in the theory of uniform approximation by analytic functions of one complex variable. Outlines of proofs and even some definitions will be given without the precision which would normally be expected. My intent is to impart some of the flavor of the subject without requiring digestion of a full meal. Fuller treatment with more precision can be found in the references cited at the end of this paper, and further references can be found in [2,5,9,10].

Let M be an open Riemann surface, that is, a non-compact connected one-dimensional complex manifold-without-boundary. For each closed subset E of M define A(E) to be the set of continuous functions defined on E which are analytic on the interior E^o of E. Consider these conditions on the pair (M,E), which may or may not be satisfied in any given instance.

(A) Every element of A(E) is the uniform limit on E of functions which are everywhere analytic on M.

(*) M* - E is connected and locally connected, where M* is the one-point compactification of M.

The meaning of (*) is that each point not in E can be joined to ∞ by a path in M which avoids E and that if the starting point is suf-

* Partially supported by NSF Grant MCS 77-01848

ficiently far away the path can be chosen to avoid as well any pre-assigned compact set. When E is compact, $M^* - E$ is automatically locally connected.

THEOREM 1. (A) implies (*).

For $M = \mathbb{C}$ this was proved by Keldyš and Lavrentiev[8] in 1939; however, it may be that Carleman and Roth were aware of it earlier (oral communications from Roth to Gauthier and from Gauthier to me). For general M this was proved by Gauthier and Hengartner [7] in 1975.

THEOREM 2. In certain situations (*) implies (A).

(a) For $M = \mathbb{C}$ and E compact this result is the celebrated theorem of Mergelyan (1951) [9]

(b) For general M and compact E this result is due to Bishop (1958) [3].

(c) For $M = \mathbb{C}$ and E arbitrary this is the famous theorem of Arakelyan (1963) [1,2,4].

(d) For general M and $E^o = \emptyset$ (so that $A(E)=C(E)$), Gauthier and Hengartner proved this result in 1973 [6,10].

(e) However, Gauthier and Hengartner provided an example in 1975 to show that it is not always true that (*) implies (A) [7].

(f) In the presence of an additional hypothesis, call it (**), (*) implies (A).

Roughly, (**) says that $\overline{E^o}$ is covered by a locally finite family of closed sets which overlap only in 'thin' sets and each of which has a neighborhood of finite genus. A thin set is, for example, one which is an interpolation set for bounded elements of $A(E)$. Theorem 2(f) subsumes parts (a) through (d); its proof uses Bishop's theorem (2b) and appears in [10]. As corollaries of (2f) we see that for surfaces of finite genus (A) is equivalent to (*) and hence is a topological invariant of the pair (M,E). However, (A) is not a topological invariant [10]. In fact,

THEOREM 3 [11]. (A) need be invariant even for a quasiconformal automorphism of M which is isotopic to the identity, even when all the mappings and homotopies are real-analytic.

Here are four examples which illustrate the existence or non-existence of uniform approximations.

EXAMPLE 1. It is not enough that M* - E merely be connected in order for (A) to hold. As an example let M be $\mathbb{C}$ and E be the positive y-axis together with this sequence of curves: γ_n= the two long sides and the top of the rectangle which has height n and base the interval $[\frac{1}{2n+1},\frac{1}{2n}]$. See figure 1.

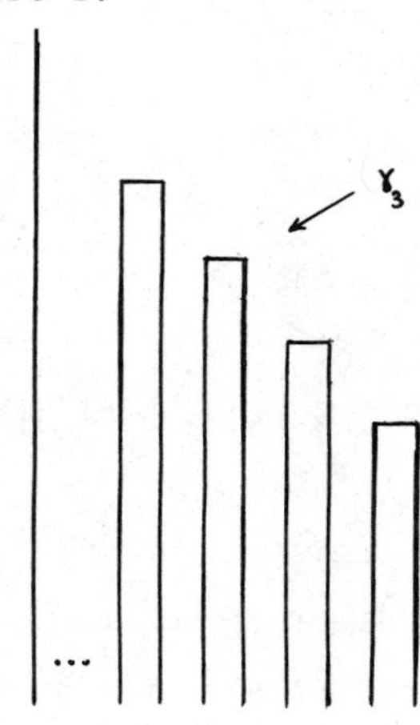

Figure 1

For a function in A(E)=C(E) which cannot be approximated choose f to be 0 on all the vertical lines in E but so that $\int_{\gamma_n} f\, dz = 2n$. If F is entire, then F is uniformly bounded on the base of the rectangles. So $\int_{\gamma_n} F\, dz = \int_{\text{rectangle}} - \int_{\text{base}} = 0 + o(1)$. Thus, if $|F-f| \leq 1$ on E we can calculate

$$\left|\int_{\gamma_n} f\, dz\right| \leq \int |F-f| + \left|\int F\right| \leq n + o(1) < 2n$$

for large n, a contradiction.

EXAMPLE 2. (*) does not imply (A). Let M be the result of taking two copies of the open unit disc D, slitting each disc along all the intervals $[\frac{2n-2}{2n-1}, \frac{2n-1}{2n}]$ for $n \geq 1$, and joining the sheets cross-wise along corresponding slits. (Roughly, if one looks at M end-on along the real axis one might 'see' .) Let $D^+ = D \cap \{\mathrm{Re}\ z \geq 0\}$ and $E = \pi^{-1}(D^+)$. See Figure 2.

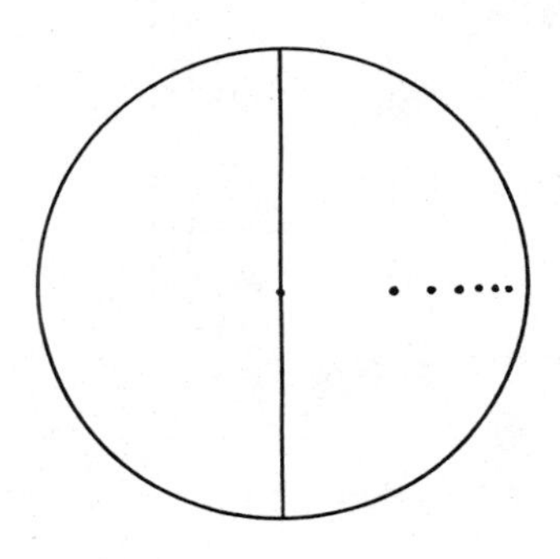

Figure 2

It is trivial that (*) holds for E in M. Define for every analytic g on E the function Δg on D^+ by the rule $\Delta g = (g(z_1) - g(z_2))^2$, where $\pi^{-1}(z) = \{z_1, z_2\}$. It is easy to verify that Δg is analytic on D^+ and vanishes at all the points $1 - \frac{1}{n}$, where z_1 and z_2 coalesce. Thus, if g is bounded on E, $\Delta g \equiv 0$ on D^+, because the Blaschke condition is violated. To obtain a non-approximable f simply choose f to be meromorphic on M with its only singularity a pole at p, where $\{p,q\} = \pi^{-1}\{-\frac{1}{2}\}$. If F is analytic on M, f - F must be unbounded on E. Otherwise, $\Delta(f-F) \equiv 0$ on D^+, hence on $D - \{-\frac{1}{2}\}$, which is absurd because f - F is analytic at q and infinite at p.

EXAMPLE 3. An illustration of Arakelyan's theorem; the methods can be generalized to apply to arbitrary closed subsets of open surfaces of finite genus. The methods for this example are due to Mergelyan [9]. Let

$$E = \{\mathrm{Re}\ z \leq 0\} \cup \{z = x, 0 \leq x \leq 1\} \cup \{|z-3| \leq 1\} \cup \{z = 5+iy, -\infty < y < \infty\}.$$ See Figure 3.

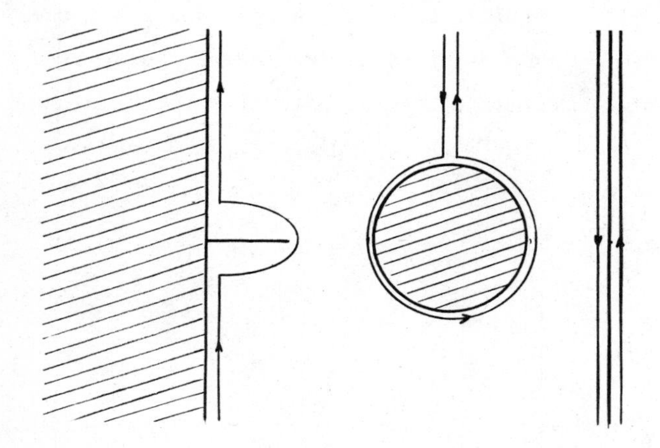

Figure 3

We approximate $f \in A(E)$ in this manner. For each disc Δ one can approximate f on $\Delta \cap E$ by a polynomial (Mergelyan theorem). Using a partition of 1 we patch these polynomials to obtain a smooth $\phi = f$ on E; $\bar{\partial}\phi = 0$ on E. Choose an open set $G \supset E$ which is close to E as measured by area, so that ∂G consists of smooth curves, and so that ∂G satisfies (*). (In Figure 3 ∂G is shown as four oriented curves.) Because ∂G satisfies (*) one can approximate ϕ very closely on ∂G by an entire function g. If ϕ is close enough to f, G is close enough to E, and g is close enough to ϕ, the following computation will succed. Let $R_n = \{|z| \leq n\}$ and $d\omega = (2\pi i)^{-1}(z-z_0)^{-1}$, where $z_0 \in E$.

$$\int_{\partial(G\cap R_n)} (\phi-g)\,d\omega = \int_{(\partial G)\cap R_n} + \int_{G\cap(\partial R_n)}$$

(1) $\|$ (4) $\|$ 0 (5) $\|$ $h(z_0)$

$$\phi(z_0)-g(z_0) - \iint_{G\cap R_n} \bar{\partial}\phi \wedge d\omega$$

$\|$

$$\iint_{E\cap R_n} + \iint_{(G-E)\cap R_n}$$

(2) $\|$ 0 (3) $\|$ 0

(1) is a consequence of Green's theorem. (2) follows from the closeness $\bar{\partial}\phi$ to 0 on E. (3) follows from the closeness of G to E. (4) follows from the closeness of g to ϕ on ∂G. (5) follows from a computation similar to (1) - (4) which shows that $\int_{G\cap\partial R_n}$ converges (as $n\to\infty$) to an entire function h. Thus $f = \phi = g + h$.

EXAMPLE 4. Theorem 3 will be illustrated by a surface M, a subset E, and a smooth isotopy $u = u_t(z) : M\times[0,1] \to M$ with the properties: u_t is for all t a quasiconformal automorphism of M, and (A) holds for $(M,u_t(E))$ if and only if $t = 0$.

Let S_λ be the strip $\{|\mathrm{Im}\, z|<\lambda\}$ and define $B(T,\lambda)(z) = \prod_{t\in T}(\phi(t)-\phi(z))\times(\phi(t)+\phi(z))^{-1}$, where $T\subset\mathbb{R}^+$ and $\phi(z) = \exp(\frac{\pi z}{2\lambda})$. When it converges $B(T,\lambda)$ is analytic on $S_{2\lambda}$, is bounded by 1 on S_λ, and has T for its zero set. One can choose $1 = a_1<<b_1<<a_2<<b_2<< \ldots$ and finite sets $T_n\subset(a_n,1+a_n)$ so that $B(\cup T_n,\lambda)$ diverges for every $\lambda>1$ and yet $p_n = B(\overset{n}{\underset{1}{\cup}} T_j,2)/B(\overset{\infty}{\underset{n+1}{\cup}} T_j,1)$ converges and is nearly 1 on the region $\{|\mathrm{Re}\, z - b_n|\le 3\}$. Slit S_2 along the disjoint intervals whose endpoints are $\cup T_n$, join two copies of this slit strip as in Example 2, and this time omit (i.e., delete) all the enpoints $\cup T_n$. The resulting M with its projection π is a two-sheeted (unbranched) covering space of $S_2 - \cup T_n$.

Put $E_\lambda = \pi^{-1}(\bar{S}_\lambda)$. If $1<\lambda<2, E_\lambda \subsetneq M$ and 0 is the only bounded analytic function on S_λ which vanishes on $\cup T_n$. As in Example 2 this implies that (M,E_λ) fails to satisfy (A). It is clear that M admits a smooth isotopy u which maps each vertical line into itself (i.e., x is the x-coordinate of $u(x+iy,t)$) and which moves E_1 smoothly through various $E_\lambda, \lambda>1$, for $t > 0$ (with $u(.,0)$ the identity map). Thus, $(M,u_t(E))$ fails to satisfy (A) for $t > 0$.

To see that (M,E_1) does satisfy (A) we use the following scheme. Because of the location and orders of the zeros and poles of p_n, the function $\pi_n = \sqrt{p_n e^{z-b_n}} \circ\pi$ is single-valued and analytic on M, is a homeomorphism on

$$\sigma_n = E_1 \cap \pi_n^{-1}\{e^{-1}<|w|<e\} = \pi^{-1}(\bar{S}_1 \cap \{|\mathrm{Re}\ z - b_n| \le 2\})$$

π_n maps the shaded regions of Figure 4a into the correspondingly shaded regions of Figure 4b.

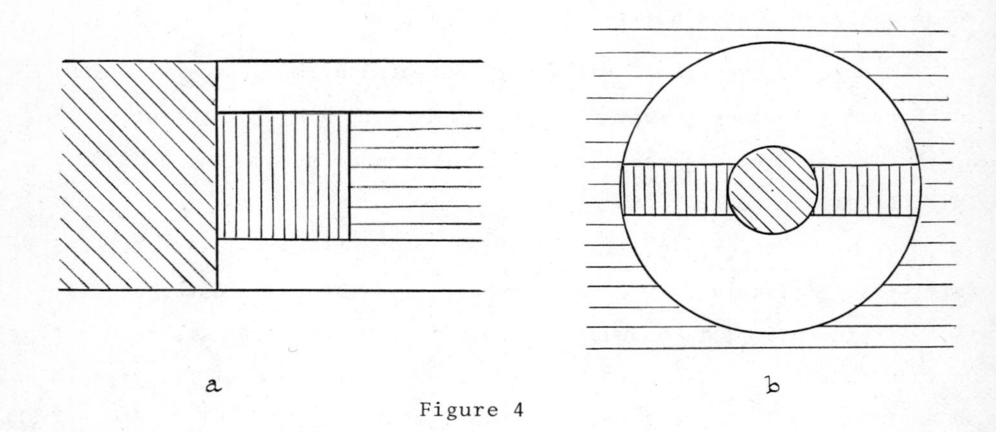

Figure 4

Given $f \in A(E)$ one can choose F analytic on M so that $F = f$ on $E' = E - \cup\sigma_n$, because $\overline{E'}$ is a locally finite disjoint union of compact sets, each of which satisfies (*). Let χ_n be 1 on σ_n and 0 off a neighborhood of σ_n. Then $g = (\sum\chi_n)(f-F) = f-F$, because $\sum\chi_n = 1$ wherever $f-F$ is not small. $g_n = \chi_n(f-F)$ has small $\bar\partial$ on E^o because $\bar\partial(f-F)=0$ on E^o and $f-F$ is small wherever $\bar\partial\chi_n \neq 0$. Because π_n separates the two sheets of σ_n, g_n can be factored as $g_n = G_n \circ \pi_n$, where G_n has compact support and $\bar\partial G_n = 0$ on the shaded portions of Figure 4b. It follows (see [9]) that G_n can be approximated reasonably well by a rational function r_n whose poles lie in the unshaded part of Figure 4b. Thus, $r_n \circ \pi_n = g_n$ on the shaded portions of Fugure 4a. The poles lie in the unshaded parts of Figure 4a; these can be eliminated by multiplying by a function which is nearly 1 on the shaded parts and which vanishes to high enough orders where it is needed to vanish. Then these analytic approximations to g_n can be added to yield an analytic function H on M which is near g on E. So $F + H = f$ on E.

I shall close now with some open questions. Denote by B(X) the set of functions which are bounded uniformly on X and consider these conditions on a pair (M,E).

(i) $A(M) \cap B(E^{o})$ contains a non-constant function.

(ii) $A(M) \cap B(E^{o})$ separates points of E.

(iii) For each point of E^{o} some member of $A(M) \cap B(E^{o})$ is a local coordinate near that point.

(iv) $A(E) \cap B(E^{o})$ contains a non-constant function.

(v) $A(E) \cap B(E^{o})$ separates points of E.

(vi) $A(E) \cap B(E^{o})$ contains a local coordinate for each point of E^{o}.

QUESTION. Is any one of these (or any combination) either necessary or sufficient for (A) (in the presence of (*), of course)? Note: (iv) is necessary, assuming $E \subsetneq M$.

QUESTION. If (A) holds for (M,E), are there arbitarily large compact K so that (A) holds for $(M, E \cup K)$?

QUESTION. If (M,E) satisfies (A) and $E' \subset E \subsetneq M$ is such that (M,E') satisfies (*), does (M,E') satisfy (A)?

REFERENCES

[1] N. U. Arakelyan, Uniform approximation on closed sets by entire functions (Russian), *Akad. Nauk SSSR. Izvestia, ser. mat.*,28 (1964), 1187-1206.

[2] N. U. Arakelyan, Approximation complexe et propriétés des fonctions analytiques, *Actes Congrès intern. Math.*, 1970, t.2, 595-600.

[3] E. Bishop, Subalgebras of functions on a Riemann surface, *Pacific J. Math.* 8 (1958), 29-50.

[4] W. H. J. Fuchs. *Théorie de l'Approximation des Fonctions d'une Variable Complexe*. Séminaire de Math. Supér. Univ. de Montréal, 1968.

[5] P. M. Gauthier, Meromorphic uniform approximation on closed subsets of open Riemann surfaces, to appear

[6] P. M. Gauthier and W. Hengartner, Approximation sur les fermés par des fonctions analytiques sur une surface de Riemann, *Comptes Ren-*

dus de l'Acad. Bulgare des Sciences (Doklady Bulgar. Akad. Nauk) 26 (1973), 731.

[7] P. M. Gauthier and W. Hengartner, Uniform approximation on close sets by functions analytic on a Riemann surface, *Approximation T ory* (Z. Ciesielski and J. Musielak, ed.), Reidel Publ., Holland, 1975, 63-70.

[8] M. V. Keldyš and M. A. Lavrentiev, Sur un problème de M. Carlema *C. R. (Doklady) Acad. Sci. USSR (N.S.)* 23 (1939), 746-748.

[9] S. N. Mergelyan, Uniform approximations to functions of a comple variable, *A.M.S. Translations, Series one,* 3 (1962), 294-391.

[10] S. Scheinberg, Uniform approximation by functions analytic on a Riemann surface, *Annals of Math.* 108 (1978).

[11] S. Scheinberg, Non-invariance of an approximation property under quasiconformal isotopy, to appear.

Department of Mathematics
University of California
Irvine